ON THE ORIGIN OF
DISEASES

Dr. Neil B. Benson, MD

ISBN: 978-1-6847-1596-1 (sc)
ISBN: 978-1-6847-1598-5 (hc)
ISBN: 978-1-6847-1597-8 (e)

Library of Congress Control Number: 2019920565

Interior Image Credit: Clint De Vos

Lulu Publishing Services rev. date: 01/24/2020

Acknowledgements

To my parents, Margaret England and the late Dr. K. I. G. (Ken) Benson, who was for many years the Assistant Deputy Health Minister of British Columbia, Canada, and became Professor Emeritus at UBC. Both my parents gave me support in my endeavors and the Scottish genetics for stubborn perseverance despite the odds, that was crucial in discovering the causes of allergies and autoimmune diseases.

To my four children; Lystra, Hugh, Geoffrey, and Philippa and their mother Helen, who all suffered along with me the financial and social penalties associated with my work. With special thanks to my son Hugh who gave up several months of his life to direct, support and enable the creation of this book.

To my brother Professor Iain T. Benson, who had an early appreciation of the importance of my book and who gave me valuable suggestions on the introduction.

To a past patient, Sandra Crashley, who very early on recognized the significance of my work. She encouraged and enabled me to speak at the CFS conference in 1995 where I first presented the information on the "abnormal bowel cascade".

To Dr. Patricia Sherwood who explained veganism to me and how to establish a healthy vegan diet.

To Dr. Xue Wang who gave me support and encouragement, and my friends Sue Footner, Christopher Wilson and Madeline King all of whom did proofreading and made suggestions.

To the Department of Botany at the University of British Columbia who gave me a wonderful education and inspired me to continue in the sciences.

To the Natural Science and Engineering Research Council of Canada (NSERC) who awarded me a five-year scholarship to undertake a PhD anywhere in Canada. This recognition encouraged me and the financial support made it possible for me to attend graduate studies. The one year I completed cemented my love of science and introduced me to Professor C. Finnegan, Dean of Science, who stimulated discussion, encouraged and rewarded creative thought and sent his students out to explore the many great minds that make up the study of the philosophy of science.

To Dr. David Hardwick, Professor Emeritus at the University of British Columbia's Department of Pathology who acted as a role model. He recognized my strong social conscience and enquiring mind and supported me through my medical training.

To Dr. Brenda Murrison, my colleague who allowed me to work as a doctor in her medical clinic despite my controversial views.

Finally, to all the generous patients who shared their experiences with me. Without them my knowledge could not have grown and this book would not exist.

Introduction

This book comes from my life's work as a medical doctor. While scientific in nature, the book has been written with the aim of appealing to both those in scientific disciplines and to the enquiring public at large. As this book has been written based on my working vocabulary, references have been kept to a minimum except where direct quotations have been made. I recognize the importance of the information contained in this book, and I am aware that there will be hesitancy and resistance from some quarters. In *The Structure of Scientific Revolutions*, Thomas Kuhn (1962) famously has pointed out that all paradigm shifts involve controversy before acceptance.

Professor Ernst Mayr, philosopher of biology, said "the essence of science is the continuing process of problem solving in the quest for an understanding of the world in which we live" (Mayr 1982). The problem solved here is the causative mechanism for chronic inflammatory disease, which includes allergy and autoimmunity. The causation is chronic infection, and this knowledge has massive implications to medicine. I have endeavored to show how this new scientific understanding of chronic disease has the capacity to change the health of our society and all societies.

Like many of you who are reading this, I have suffered a lifetime (in my case sixty-two years) of suboptimal health. This book aims to give you an overview of what I have learned about how health works. This is based on more than thirty years of clinical work and personal research in medicine. All knowledge evolves through time, and the information presented here has made me an innovative, successful, and a personally healthier medical doctor. Application of this knowledge has improved the health of thousands of patients, most of whom had exhausted the bounds of traditional medicine in their search for relief and ease of their suffering. This book discusses diet as a fundamental aspect of our health. While many people know this, they do not know *how* and *why* diet affects their health. We will deal with how the gut works and how changes to the gut described as the "abnormal bowel cascade" influences, and in many cases causes, chronic disease. The aim of this book is to explain poor health causes rather than addressing symptoms. This is a bold, provocative approach that challenges the symptomatic treatment focus that underlies most of modern medicine. The book goes further by providing you with the knowledge and the tools to improve your own health.

This book will help the following disorders, which include the areas of allergy, autoimmune disease, immune dysfunction, nutritional deficiencies, mental health, chronic infection, and even cancer. The framework of knowledge that surrounds these disorders will allow you to start taking control of your health. Some of you will have disease at the mild end of the illness spectrum, while others will be at the extreme end. The commitment for individuals suffering at either end of the spectrum will be very different. If you find yourself at the extreme end and have been ill for a long time, the process of returning to health will be more difficult, requiring more commitment on your part. At the mild end you will experience improved health with only minor adjustments to your life. Whichever end you may find yourself on, I hope you have the perseverance to take advantage of the knowledge presented here. This knowledge has been gained thanks to many factors—the benefits of a strong science background, medical training, the observations of many medical and scientific pioneers (I stand on the shoulders of those who have gone before), the innumerable generous patients who have shared their stories and experiences, my own hard work, and luck.

As medicine is a complicated field in which many symptoms are common to a variety of diseases, you should always seek individualized health advice from a trained health professional. The risks associated with self-diagnosis can drastically impact your health, so this book has been written to inform, not to advise on your unique complicated problems.

As a disclosure, I have received no money for this work from the pharmaceutical industry or from any government, including Canada, New Zealand, or Australia, apart from patient subsidies that all GPs receive. No medical funding organization has contributed in any way financially or otherwise to my work or this book.

My Story
Physician, Heal Thyself

Prior to completing my medical degree in 1988 at the University of British Columbia, Vancouver, Canada, I had obtained two degrees: Bachelor of Science in Botany (B.Sc.) and a Bachelor of Science in Agriculture (B.Sc. Agr.) with honors in plant pathology and had completed a year of a PhD in plant ecology. This education provided me with knowledge of fungal organisms, food production methods, food chemistry, nutrition, information on botanical groupings, and an ecological way of thinking. This unconventional premedical background coupled with my own health issues proved to be ideal for unravelling the cause of allergies. Who would have thought?

Allergies are incredibly common today, and I believe they now affect at least 50 percent of the population. When I grew up some sixty-plus years ago allergies were rarely talked about. Grass, bee/wasp allergies, and pet allergies were only occasionally seen. An allergy to penicillin was recognized, and a peanut allergy was a medical curiosity but something you would be unlikely to hear about firsthand. Atopy (allergic disorders) that includes asthma, eczema, and sinus problems were relatively uncommon.

Uncommon they may have been, but unfortunately for me, I had asthma, eczema, and recurrent sinusitis, no doubt caused by, as my mother put it, my "growing up on antibiotics." While antibiotics are "wonder drugs" to treat individual infections, they have negative side effects. The negative side effects were not appreciated when I was young, and antibiotics were commonly used for minor ailments. As I was a "sickly child," antibiotics become a regular part of my life.

The frequent sinus infections I suffered from in my childhood had become poorly controlled by antibiotics, and this resulted in my being referred to an ear, nose, and throat (ENT) surgeon. Under his direction I had an initial surgery to remove my adenoids and tonsils and then a second sinus surgery to scrape my nose and put "windows" into the sinuses. Each surgery gave me mild temporary relief, but ultimately the operations were deemed failures. Following these failed operations, I was skin tested for allergies. Skin testing for me yielded no helpful information, and I subsequently believed allergy testing to be a useless tool.

About the age of ten, the ENT surgeon suggested further surgery, but I declined this, and I also refused to take any more antibiotics. Following this decision my health slowly improved, and my teenage years were relatively healthy. When I was twenty-three and attending McGill University, I became infected by impetigo (a bacterial skin infection also known as school sores). No doubt my poor diet at the time contributed to my getting the infection. The impetigo was treated with a broad-spectrum antibiotic, and following this, I found my sinus troubles slowly returned. At the age of thirty, my sinus troubles no longer responded to steroid nasal sprays. The now complete nasal obstruction caused me to reluctantly seek the help of an ENT surgeon. The first ENT surgeon felt I would be better served by his colleague, who specialized in the state of the art minimally invasive endoscopic surgery. This second ENT surgeon told me I would need major surgery, which would require me to have seven weeks postop recuperation. At that time being off work for seven weeks was an impossibility. Not only was I a young GP with a fledgling medical practice, but I had three small children and a large mortgage. The specialist further went on to say that I would feel much better after the operation but that the operation would likely have to be repeated "every couple of years" on an ongoing basis! I was not happy to hear this and asked him what was causing the complete nasal obstruction. He said it was allergies. Medical school taught me essentially nothing about allergies and falsely confirmed my preconception that allergies were unimportant. Surely, if allergies were important, I would have learned about them there, or so I believed!

At that time allergists were rare in British Columbia, Canada. Allergists were only available in the largest city—Vancouver. Vancouver was several hours travel from where I was living in Victoria, and allergists had long waiting lists. Both the waiting time and the travel distance involved were very inconvenient for me, so I decided it was time I learned about these "unimportant" allergies.

I obtained an allergy skin test kit from a Canadian allergy company. The basic test kit was for environmental allergies only and not for food. I was happy with this, as I wrongly believed that I did not have food allergies. Besides, how could food affect my sinuses? With my newly acquired allergy test kit, I was my own guinea pig. I discovered a major house dust mite allergy and several mild environmental allergies. Armed with the knowledge of the house dust mite allergy, I proceeded to read everything I could on the subject. I read about house dust mite ecology and changed my environment, getting a new bed and bedding and removing the old bedroom carpet. These changes resulted in a mild symptomatic improvement. That was all I needed, and I was now hooked on allergy and wanted to know all about it.

After this initial success, it was easy for me to test my own patients for allergies. From testing my own patients, word quickly spread in the community that there was a doctor interested in allergy. New allergy patients arrived, and this gave me further experiences and showed me the limitations and benefits of skin-prick testing. I was steadily gaining knowledge in allergies.

Early in 1993, my family and I moved to Dunedin, New Zealand. At that time there were no official allergy specialists in New Zealand. Doctors looked at allergies from other disciplines, including ENT and general practice. Dunedin had nobody working in allergy, and seeing a useful niche, I set up an allergy clinic in Dunedin as a GP specializing in allergy.

Living in Dunedin, where the medical school was, gave me the benefits of the University of Otago medical library. I devoured the journals of the day and was able to read the research into allergy going back many decades. While much of the information was confusing and contradictory, there were threads of logic and studies that even dealt with the area of food allergy. This influx of academic material coupled with the stories that were shared by unwell and generous patients started to give me the confidence to branch out into the area of food allergies. I even moved into the nonmainstream areas of diet, gut flora, and *Candida albicans*. The process of gaining knowledge was not straight-forward. It was a lot of study, guesses, failures, and some eureka moments. Analysis of the allergy literature was helped by my strong science background, and my ecological training was beneficial in understanding the internal ecosystem of the gut.

Not long after opening my allergy practice in Dunedin, a patient presented me with a copy of Dr. William G. Crook's book *The Yeast Connection*. This book changed the way I viewed Candida albicans (much more on this later) and started me on the journey to discover the causes of allergies. This book, along with many others, gave me pieces of the story and allowed me to test theories with both myself and other patients. I discovered that *C. albicans* was an issue with my own health and that treating it made my health improve. It was a time of confusing discoveries. I still remember the moment that I looked back at the dietary history of patients and realized the importance of green vegetables. I noticed that people who ate a diet high in green vegetables were much less likely to have allergies. It was then that I discovered the importance of green vegetables for feeding the gut flora. As my knowledge and confidence grew, theories began to emerge, and I coined the expression the "abnormal bowel cascade." It was at this time I deduced how allergies are caused, and I presented my theories on allergies to an international symposium on chronic fatigue syndrome (CFS), held in Dunedin. I had been asked to be a speaker at the conference, and it was here that I first presented the theory of the abnormal bowel cascade. This theory nicely explained the many symptoms present in CFS patients.

CFS has an amazing diversity of symptoms. Most doctors at the time felt that CFS patients were "nutty." If you listened carefully to them, however, you would see that most of these patients wanted to be well and that they were suffering a disability that their doctor was not able to treat. I was having some limited success with these patients, and before long I was inundated by patients with this disorder. I perhaps found it easier than most doctors to listen to and believe their stories, as I too was suffering from allergies and fatigue.

Shortly after discovering how allergies were caused, I was in contact with a pharmaceutical company as I was interested in some possible joint ventures. Our project did not eventuate and in September 1994, I received a letter back from the CEO of the pharmaceutical company. I recommend you read this letter, as it is enlightening when it says "Without being too cynical, I think you are aware that the pharmaceutical industry, and our two major client groups, i.e. the conventional medical practitioner and the retail pharmacist are basically not interested in the cure of illness or disease". As you will appreciate, I have redacted the names for obvious reasons.

As my knowledge increased, my success with patient outcomes also improved. My reputation was growing, and people came to see me from all over New Zealand, with some coming from as far away as Australia. I was booked up weeks in advance, and I loved my work. I look back fondly upon this time, until the day that the Medical Council of New Zealand charged me with disgraceful conduct.

It was four o'clock on Christmas Eve when the courier dropped off the package from the Medical Council of New Zealand (MCNZ). I signed for the parcel, and the courier retired, leaving me curious to see what the Christmas gift was. To my horror, there was a complaint and a charge from the Medical Council that was at the highest level of the medical misconduct scale. This charge was totally inconsistent with the complaint, which involved a relatively minor skin rash. The charge was vague, stating that it was the "totality of my care" that was in question. This is not a real charge, and it meant that the MCNZ were going on a fishing trip to find something more significant to charge me with. I knew that I had created waves with some doctors, but I was naïve. I felt that the medical world would welcome the innovative approach I was developing, but I was wrong.

To cut a long story short, it took me three years to completely clear my name and be found innocent of all charges. The process was emotionally and financially exhausting and was anything but fair. My health, finances, and family all suffered. It destroyed my allergy practice and the allergy products business I had developed.

For the first two years of the three-year hearing process, I was unemployed. This was due in part to the many delays and postponements and to the generally unpredictable nature of the hearing process. A change in registration was also enacted, and this prevented GPs, like me, from continuing to work independently as allergists. Ultimately, I found that when you are unsure of the process, it is very difficult to plan. Having an ongoing charge against me in New Zealand also meant I could not return to Canada to work.

After two years and with the case still being unresolved, I eventually, reluctantly, returned to general practice. I could no longer financially or emotionally afford the fight, which had already taken a massive toll. I continued my interest in the field of allergies as a GP, less visible, under the radar. Eventually, after three years, I was cleared by the MCNZ of all wrongdoing. This dreadful lose/lose encounter with the MCNZ showed me that dark forces can be present in medicine and health care. My unorthodox approach to allergy, though scientific, has been ignored by almost all my colleagues. I have found you cannot teach people who are not open to learning, but you can lay out a blueprint for those who wish to follow. This book is the blueprint.

Why now for the blueprint? Science continues to develop, and talk of the gut microbiome is now an accepted mainstream field of discussion. Nutritional knowledge has also changed considerably since the 1990s. Once when I was presenting at a public meeting a senior nutritionist stood up and called me a "quack". This was because I stated that green vegetables were important in the diet of children; she disagreed and felt fruit was all that was needed. The role of green vegetables is now recognized as being important for the gut flora and for human health, so time proved me correct.

Allergy causation is also finally being looked at, and as I am coming to the end of my medical career, I feel it is important to share my knowledge formally on this subject. This book will stimulate and direct scientific research along more fruitful lines, wasting less time and money. I also know this book contains medical knowledge that improved my health and the health of many others. I am sure that in the future more people will also benefit. Finally, I would like this book to be my small contribution to the world.

The Abnormal Bowel Cascade

To understand health, we are starting at the "guts of the issue." No talk of health can be complete without mentioning the bowel, which is intimately related to most aspects of human health. Hippocrates, who is considered the father of medicine, is quoted as saying "all disease begins in the gut." This statement was largely correct twenty-four hundred years ago, and it remains true today.

For clarity, we will discuss the body in a simplified way. This most basic plan is seen as a tube surrounded by the body. The tube starts at the mouth and ends at the anus. This tube, which we know as the gut, is about 8.5 meters long, and it has four hundred square meters of surface area. It is covered by a mucous membrane that is one cell thick and is supported by subcutaneous tissues in which blood vessels and nerves run and body defense cells reside. These four hundred square meters of surface area are very large when compared to the area of the skin, which is only two square meters.

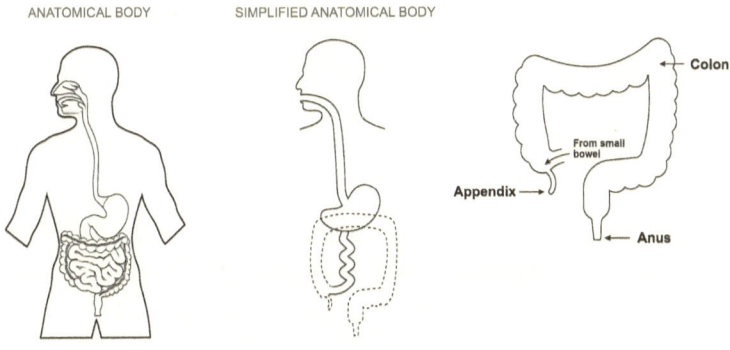

ANATOMICAL BODY   SIMPLIFIED ANATOMICAL BODY

Colon

From small bowel

Appendix

Anus

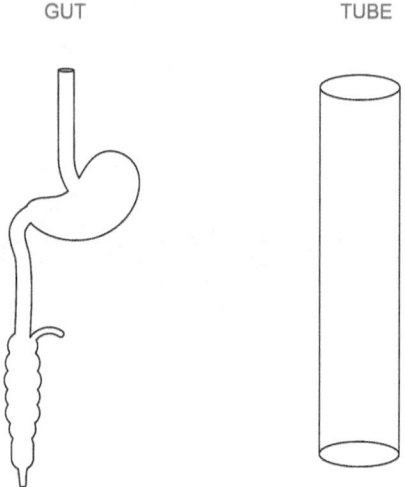

If we were to straighten up the gut from the mouth to the anus and open the muscular sphincters, we would be able to see down and through the gut to what was on the other side. We would be looking down an 8.5-meter tube. What is apparent is that what is inside the gut is not actually inside our bodies. To use an analogy, the body is shaped like a donut. The hole inside the donut is not a part of the donut any more than the gut lumen is a part of the human body. We literally are a hollow tube.

SIMPLIFIED BODY   SIMPLIFIED BODY (DONUT)

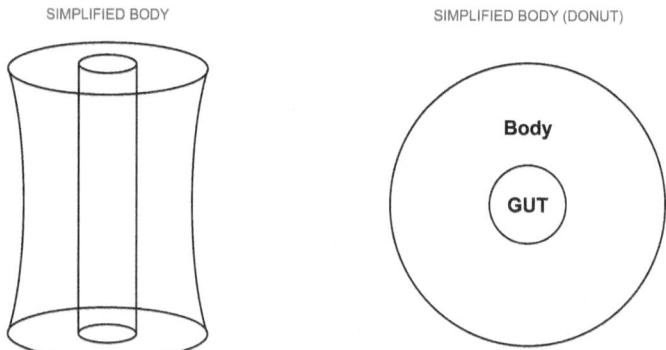

This gut tube is divided by four sphincters (muscular valves), which in turn are divided into three different regions that have different jobs or functions. The stomach is predominantly for storage and the small bowel digests and absorbs nutrients while the large bowel hosts most of the beneficial gut flora and reabsorbs fluids prior to defecation.

Gut Functions

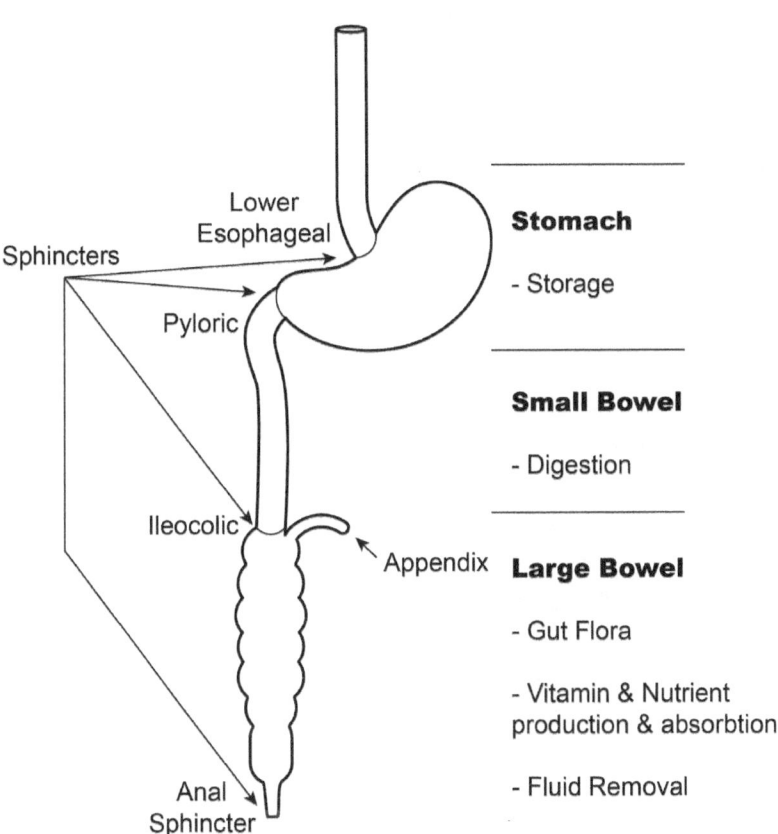

Starting at the top of the tube is the mouth. The mouth chews food and breaks it into small, digestible pieces that are then transported to the stomach, where the food is stored. From the stomach food is slowly moved on to the small bowel, where it is broken down into its constituent parts in a process known as digestion.

It is in the small bowel that most food breakdown occurs. Proteins, carbohydrates, fats, and oils are digested into their constituent smaller parts: amino acids, sugars, triglycerides, cholesterol, and so on. These are then absorbed across the mucosa of the small bowel into the body. Other breakdown products of the food result in the freeing up of vitamins and minerals, which are also absorbed from the gut. This picture is then completed by saying that the 'liquidy' remnants pass along to the large colon, where fluids are reabsorbed. The remnants are then packaged, with a sausage-like "poo" being the result. In a normal bowel the passing of a feces occurs once a day.

We are all aware that we should eat a good diet. What constitutes a good diet depends on who you ask. Some dietary recommendations are much better than others. What a good diet must have is the building blocks that allow for repair, possibly growth, and sufficient calories to meet the energy requirements of the individual. These needs will vary with a person's health, age, and activity level. This simplified view is how many experts refer to diet, but this view is incomplete.

The problem with this picture is that it misses the importance of the gut flora, a diverse group of microscopic organisms that make their home in our bowel. The gut flora is also known as the microbiome. It is made up of more than five hundred species of bacterial organisms and perhaps thousands of other types of organisms. Most of these bacteria reside in the large bowel, where our body is in a beneficial (symbiotic) relationship with them. Until recently, the role of bacteria in the gut was largely ignored by medicine.

So, what is the role of the gut flora that is predominantly found in the large bowel? One role relates to vitamins. Our body is dependent on a range of vitamins that are involved in a support role within our bodies. Each vitamin has a unique function in the body, but in general they are all important for cell metabolism. Vitamins are an interesting area that will only be touched on in this book. Here it will suffice to say that the B vitamin group, which includes niacin, thiamin, folate, and the vitamin biotin, have all been found to be produced by one or more bacteria of the gut flora. Bifidobacterium, bacillus, and many other species and strains of bacteria have a major role to play in producing vitamins. These nutrients can then be absorbed from the large bowel. The gut flora has a role in providing these vitamins rather than the vitamins simply being obtained from the food we eat.

Gut bacteria produce bacteriocins and other agents that act like natural antibiotics. These "natural antibiotics" help us fight infections both in the gut and elsewhere in the body.

The gut flora also competes with organisms that would like to live in the gut. If the beneficial gut flora were not present, other, less beneficial or even pathogenic organisms would settle in. This pathogenic group of organisms includes other bacteria, protozoa, and fungal organisms. The gut bacteria therefore have nutritional and protective roles in the human host. Over the last several years the gut flora has been found to produce neurotransmitters that affect the way we feel, making us less anxious and happier. The human host clearly benefits from the gut flora. Given the benefit to us, an obvious question needs to be asked: How can we help the gut flora?

The gut flora is greatly helped by what we eat. While most of the food components are absorbed in the small bowel, some food remnants are not broken down in the small bowel. These are known as fiber. Fiber comes in two types: soluble and insoluble.

Insoluble fiber is found in bran and many other foods. The outside of a corn kernel is a very good example. Corn kernels, if not chewed and then swallowed, are little changed by the process of digestion. The human gut flora cannot break down the corn casing. You may have even noticed some corn kernels being passed undigested in your feces. The gut flora, by definition, cannot break down insoluble fiber. Insoluble fiber is not beneficial as food for us or the gut flora. It does give us a more formed stool, and it helps to move the colonic contents by giving peristalsis something to push against.

It is when we eat soluble fiber that we are feeding and keeping our gut flora healthy. Soluble fiber is also known as a prebiotic (something that feeds bacteria). It is found predominantly in green vegetables, particularly the leafy and stem types. Broccoli, cabbage, spinach and, yes, cauliflower can be included in this; all provide soluble fiber. The "fruit" types of vegetables such as capsicum, cucumber, and tomato have less soluble fiber, but they are still very good for us. We should eat lots of green vegetables and a range of green and colored vegetables. Diversity in foods gives a range of nutrients and encourages diversity in the gut flora, which, in turn, helps us.

To recap: The gut is responsible for the breakdown and absorption of food in a process known as digestion. Digestion allows us to obtain the building blocks and energy resources that our bodies need. The gut also houses the beneficial bacteria. These bacteria are known as the gut flora or gut microbiome. The gut flora is greatly affected by our diet. The adage that "you are what you eat" correctly identifies that our diet gives us the building blocks and energy we need and affects the health of our gut flora. We feed the gut flora by eating soluble fiber, and the gut flora has a profound effect on our overall health, both physical and mental. While insoluble fiber has some benefits to our health, it does not feed the good gut flora. Soluble fiber found mostly in green vegetables feeds the gut flora.

How We Should Eat

Let food be thy medicine and medicine be thy food.
—Hippocrates

Our ancestors were not as lucky as we are today and did not have the availability of inexpensive quality foods provided by modern agriculture. Most traditional diets are high in plant fiber and protein (the so-called paleo- or caveman-type diets). Diets like this are probably what mankind is meant to consume.

With the advent of modern agriculture, we have seen a change in the cost of food. Modern grain products are high in carbohydrates and low in soluble fiber. Grains are easy to grow and store, and because of these benefits they have fallen the most in price. The low cost of grains has encouraged a change in most cultures to a diet high in carbohydrates and low in soluble fiber. All too often modern food production has further processed foods, robbing them of much of the prior nutrient content. White flour gives us little nutritional benefit apart from calories. These empty calories are detrimental to most people's health. We have moved away from our traditional plant-based diet to a diet based on high-carbohydrate grains. This change has negatively affected the gut flora and our health.

What is a healthy diet? There are two main considerations. We need to grow and maintain our bodies, and we need to maintain the gut flora. Fortunately, this can be diagrammed quite simply by looking at the plate of your main meal of the day (your largest meal). This should be half green vegetables (to give soluble fiber and some vitamins and minerals); one quarter protein, such as meat or beans; and up to one quarter carbohydrate.

The carbohydrate quadrant should include an orange vegetable such as carrot, pumpkin, or sweet potato. Orange vegetables are useful, as they contain the nutrient B carotene, which our body converts to two molecules of vitamin A.

Healthy Plate Diet

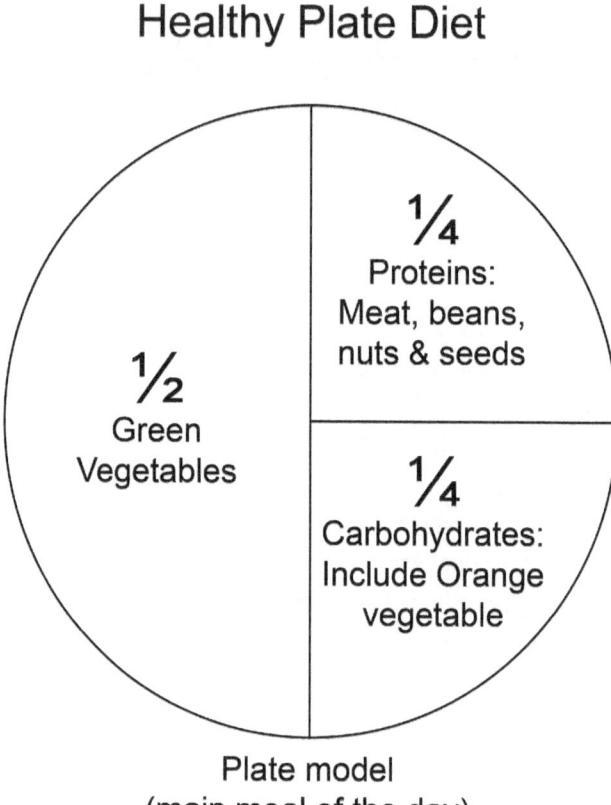

Plate model
(main meal of the day)

Vegans and vegetarians differ from meat eaters only by what is in the protein quarter. Vegans should eat beans, nuts, and seeds, while vegetarians can add eggs and dairy products. Meat eaters have these options and more. Protein for them also includes beef, lamb, pork, chicken, fish, or other animal proteins. It is possible to be a healthy vegan if you follow this simple dietary advice and remember to eat enough protein in the form of beans, nuts, and seeds. All diets, vegan to meat eaters alike, should try to limit the carbohydrates to a quarter or less of the plate.

Dr. Neil B. Benson, MD

Fruit should be limited to two pieces a day. Each piece should be no bigger than the size of the consumer's fist. If you are feeding a child, remember the amount of fruit is the size of two of their fists not yours. Less sugary fruit such as berries are preferable to high sugary fruit such as watermelon and pineapple. Berries give us the nutrients and vitamins without the higher sugar content. If the gut flora is healthy, the above dietary approach will promote and maintain a healthy gut flora over time.

The gut flora may be damaged if the diet has been poor, or it has been poisoned by antibiotics. The damage varies with the antibiotic taken, the length of antibiotic therapy, and how the antibiotic was administered. The combination of amoxicillin and clavulanic acid constitutes a very broad-spectrum antibiotic. Taking this antibiotic is particularly detrimental to the gut flora. This antibiotic is very useful in some circumstances (diverticulitis), but it is too strong for most infections, and the antibiotic should be used in a restricted manner. Narrow-spectrum antibiotics should be used whenever possible, as they can be directed at the target organism while minimizing damage to the gut flora.

Science is finding that following some antibiotics the gut flora does not necessarily reestablish as quickly as we once thought. Overgrowth with resistant varieties of bacteria, protozoa, and fungi sometimes prevents proper reestablishment of the beneficial gut flora. Broad-spectrum antibiotics change the gut flora, and these changes can persist months to years after antibiotic use. When the diet is poor, the gut flora is unlikely to recover. While antibiotics are very useful medications, they should be used wisely and prescribed carefully. Antibiotics do not give health; rather, they are a tool that can be used to rid the body of a specific problem when it occurs. Antibiotics are "magic bullets" when directed appropriately but are damaging when used indiscriminately.

Antibiotics: The 'Magic Bullet' or 'Nuke' of the gut

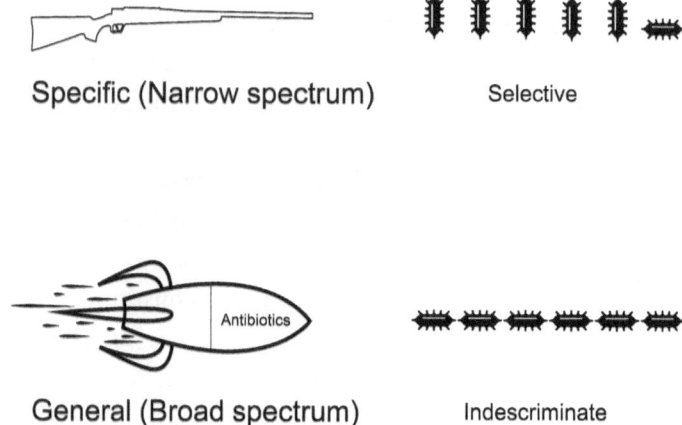

Specific (Narrow spectrum)   Selective

General (Broad spectrum)   Indescriminate

Until recently the gut flora was hard to study, and subsequently it has been poorly understood. Apart from the role of breaking down food and absorbing nutrients, little was known of the ecology of the gut and the role in human health. With the advent of DNA sequencing and other advances, our understanding has changed considerably. Many companies are now looking at this area and I believe that benefits will come from this research.

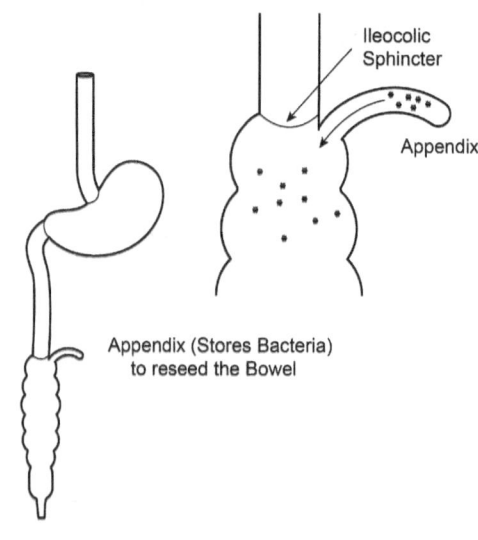

Ileocolic Sphincter

Appendix

Appendix (Stores Bacteria) to reseed the Bowel

When I was a medical student, we were taught that the appendix, which is a part of the large bowel, was essentially useless—a vestigial evolutionary remnant of little significance. We were told that the appendix had some minor immunological role but that it was something we could easily live without. We now know the appendix has an important role to play. The appendix is a bag about the size of your little finger that contains gut bacteria. It is found at the juncture of the small bowel and the large bowel. The appendix bacteria are ideally located to seed the colon when the essentially sterile digested remnants are received from the small bowel. While we can live without the appendix, it is particularly helpful in cases of diarrhea. Should the gut flora be flushed out by diarrhea, then the bacterial store in the appendix, which does not get washed out, reseeds the gut flora. The appendix is not useless but rather acts as a storage bag keeping the bacteria in our gut replenished and healthy. The bacteria in the appendix can also be damaged by antibiotics. This area of appendix health needs to be studied, as I suspect a poorly colonized appendix will present ongoing problems to the host. Can we change the appendix flora? For budding scientists there is a PhD thesis here waiting to be written and money to be made for the appropriate entrepreneur!

The Gut as a Barrier to Infection

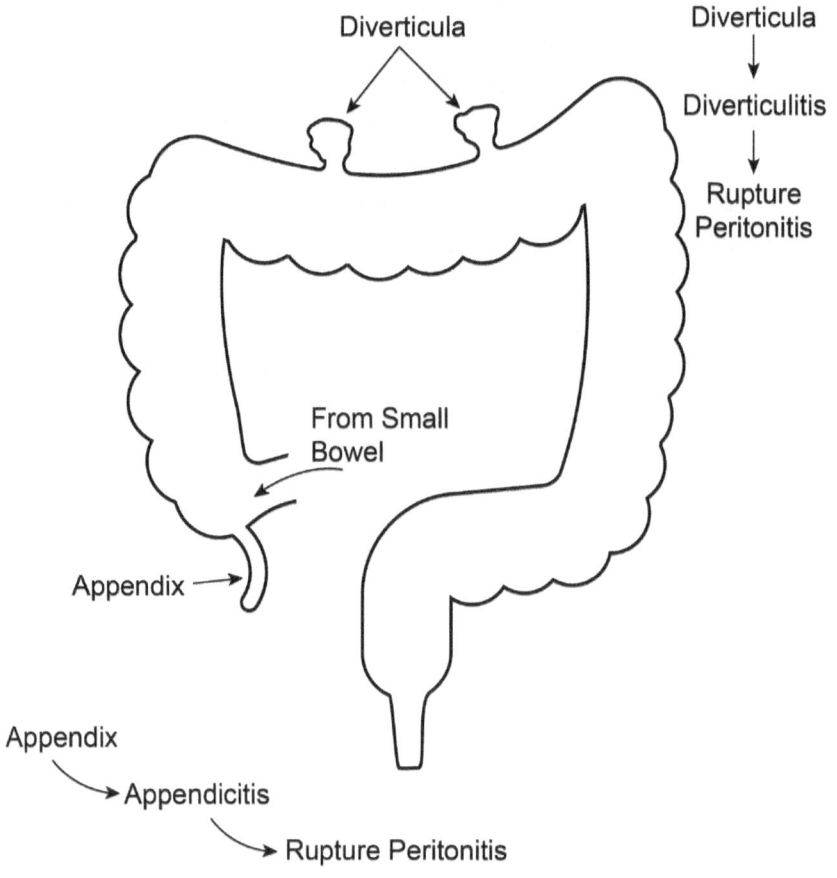

While many of the gut flora bacteria are beneficial when contained within the gut, these same species if escaping out of the gut and into the body may kill us. Appendicitis and diverticulitis are both contained infection types known as abscesses. Appendicitis is in the appendix and diverticulitis is in diverticula of the large bowel. Both can rupture, leading to an uncontrolled spreading infection known as peritonitis. Left untreated, peritonitis often leads to death. To avoid death the body has developed protective measures in the form of barriers to control and prevent the gut flora from entering the body and harming us.

How the Gut Protects the Body

The human body has several barriers to prevent invasion of the gut flora into our bodies. Starting in the mouth we have saliva, which is a mixture of water, mucous, some proteins, and enzymes. Saliva aids swallowing and washes food and organisms from the mouth to the stomach. This steady stream of mucous moves bacteria and other organisms along the gut wall, making their establishment difficult. Mucous is helped by movements of the gut wall muscles, which is known as peristalsis. Peristalsis squeezes food through the gut, and the waves of contraction look very similar to the movements of a worm. Peristalsis acts to move food throughout the gut, from the mouth to the anus.

Peristalsis

From mouth

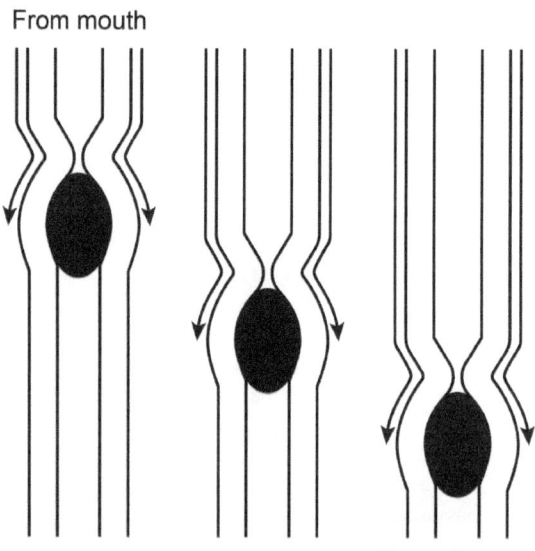

Towards anus

The movement of food from the mouth to the stomach is quick, and the flushing activity of the saliva results in little opportunity for the establishment of organisms. Food expands the stomach, and food is stored here. The stomach expansion enables us to feast on food when it is available with the luxury of being able to process the food later, when it is more convenient and can be done more efficiently.

As a storage bag, the stomach is at risk of infection. To minimize this possibility, food is held in the stomach for as short a time as possible. The risk of infection is greatly reduced in the stomach by the production of hydrochloric acid. Hydrochloric acid allows the empty stomach to sterilize itself between meals. The empty stomach is highly acidic (pH 3). This acidic environment presents a challenge to organisms, and few survive. The stomach acid also helps to break down and sterilize ingested foods.

Food stored in the stomach is now moved on to the small bowel, where it is digested. Digestion uses enzymes to break down all foods (except fiber) into their constituent parts. These highly destructive digestive enzymes not only break down foods; they kill most organisms, making the small bowel nearly sterile. This is a good thing, as we do not want organisms competing with us for our food. We have worked hard to obtain these calories and nutrients, and we do not want competition for them. Few bacteria are found in the first part of the small bowel and this number increases toward the end of the small bowel. In healthy individuals the number of organisms at the terminal end of the small bowel is still very few.

Most of the nutrients and minerals are absorbed in the small bowel. The liquid solution that now remains should be low in nutrients and hopefully rich in fiber. This solution is then moved from the small bowel into the large bowel. At the joining of the small and large bowel we find the appendix. The appendix inoculates the fiber-rich fluids with bacteria, which quickly become colonized and form our colonic gut flora. Fluids are withdrawn from the gut along with some vitamins and other compounds, and the remnant waste products are now packaged into feces. In the colon our gut flora produces the multitude of products; vitamins, bacteriocins, and neurotransmitters help our nutrition, immunity, and mental health.

Interspecific competition exists between different organisms within the gut. Beneficial bacteria and pathogens compete for food and space. Beneficial gut flora bacteria compete against and inhibit the growth of pathogens. If we have reduced beneficial bacteria, because of a poor diet or toxic agents such as antibiotics, pathogens overgrow, and our health is negatively impacted.

To recap: The gut is a tube compartmentalized by sphincters into areas that have different roles. Physical barriers protect the body; these barriers include mucous, peristaltic movement, low pH (acidic) environments, digestive enzymes and biological competition. Gut flora bacteria compete with and restrict the growth of pathogens. Vitamins and other compounds produced by the gut flora in the colon positively affect our health.

Cellular Defenses and What We Know as Immunity

We have now moved from the physical barriers of mucous, movement, and hostile environment (provided by enzymes, pH, and competition), to cellular defenses. This area is more formally known as immunity. Immunity is "the ability to resist a particular infection or toxin by the action of specific antibodies or sensitized cells" (Oxford English dictionary).

Our primary cellular defenses rely on cells that are generalists. These generalist cells respond to all invaders by acting as soldiers and cleaners. We know these cells as phagocytes, and more specifically they are called macrophages, monocytes, eosinophils, and neutrophils.

Phagocytes
(General Cellular Defenses)

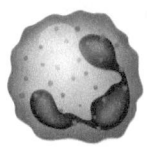
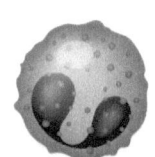

Neutrophils   Eosinophils   Monocytes and Macrophages

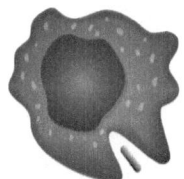

Process of Phagocytosis

OVERVIEW OF PHAGOCYTOSIS

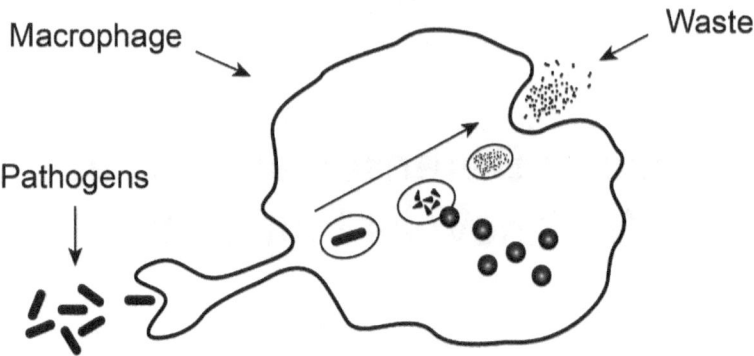

The phagocytes envelop bacteria, viruses, and a range of foreign cells and proteins much like the amoeba you probably learned about in high school. While some of the phagocytic cells are more efficient at identifying and killing certain types of infections, they are not able to tackle all infections. They lack the capacity to kill infections that have become intracellular, and they are unable to phagocytize large organisms or bacteria and organisms that have a resistant coating. For instance, some bacteria and fungi have a capsule or surface that is very difficult to break down. For these specialized tasks the body has other defense mechanisms that make up the subject of immunity. Immunity is a large and complex subject; a simplified version will be discussed on the following pages.

Immunity
Specific Induced Response

Lymphocyte types and functions

Three types of Lymphocytes

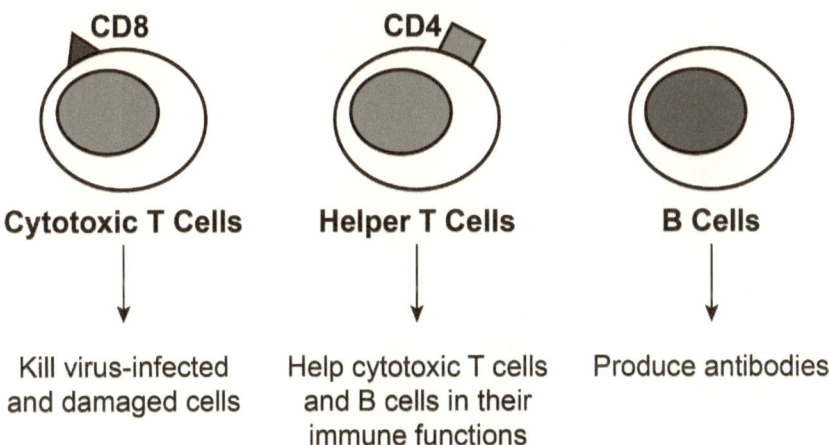

Cytotoxic T Cells	Helper T Cells	B Cells
Kill virus-infected and damaged cells	Help cytotoxic T cells and B cells in their immune functions	Produce antibodies

For specialized defensive tasks, our body has cells known as lymphocytes, and they have a memory and recall previous encounters with foreign proteins. Lymphocytes are of two classes: T cells or B cells. T lymphocytes kill intracellular infections and do other roles, while B lymphocytes produce an array of proteins known as immunoglobulins. The main immunoglobulins are known as IgA, IgM, IgG, and IgE. Of these, IgM is involved only with an initial response to infection while the other three are created in subsequent exposures. Antibodies are of different types because they perform different defensive roles. IgA is typically related to mucous membranes, be it in the gut, lungs, or eyes, while IgG performs a variety of roles, one of which includes setting off the complement system. The complement system is a group of nine proteins that fit together in a ring shape. When deployed, the complement system directly attacks targets, such as infected cells or invading organisms, by punching a hole in them. IgG also coats organisms making them malfunction and making it easier for phagocytes to identify and envelop them. The final antibody, and the most common antibody found in parasite defense and the allergic response, is IgE.

Generalized Antibody

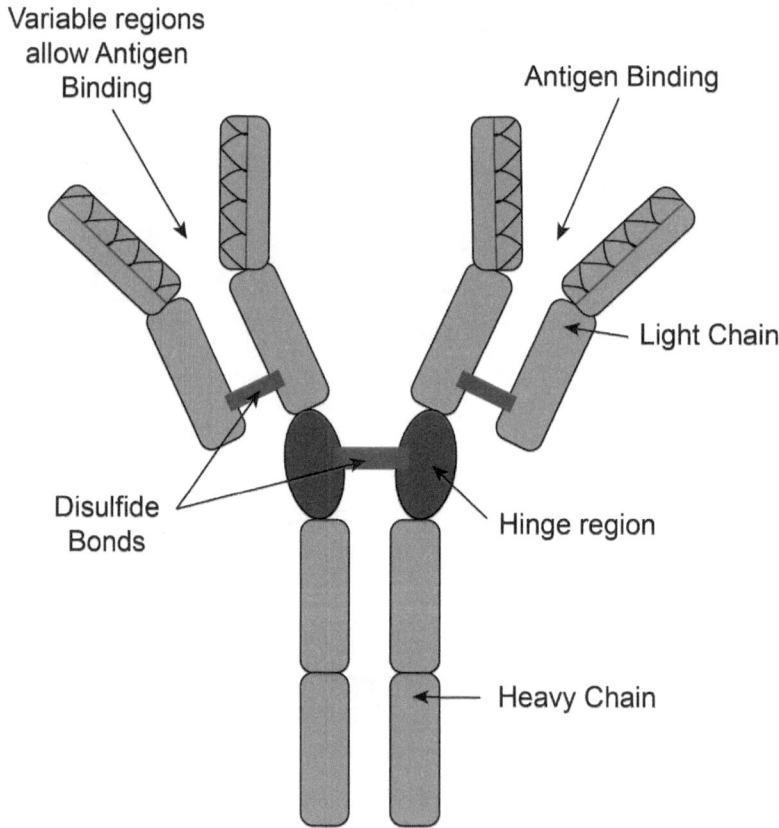

Before talking further about the types of antibodies and their roles, we must first talk about the structure of the gut wall.

Dr. Neil B. Benson, MD

Gut Wall
Physical Structure and Related Functions

Components of the Intestinal Barrier

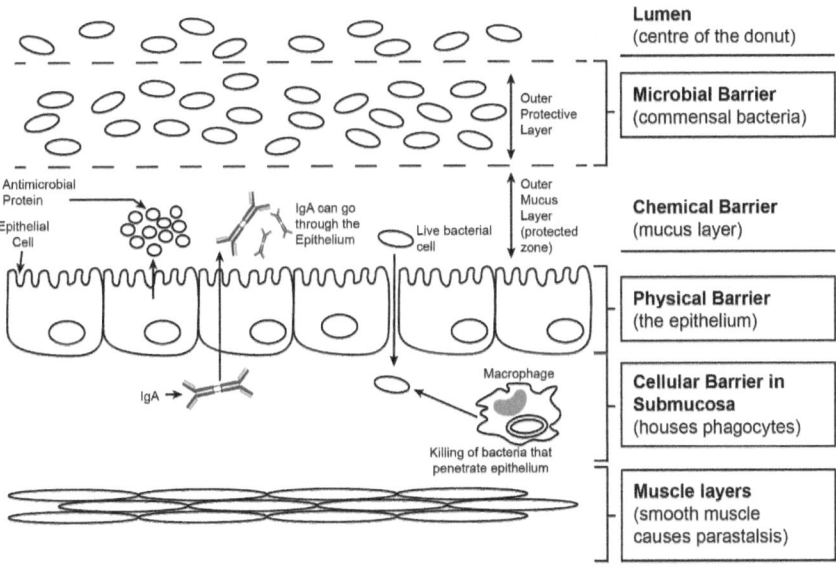

The epithelium is one cell thick and has "tight junctions" between the cells that normally prevent the passage of all large proteins. The area below the epithelium is called the submucosa. Phagocytic cells, blood vessels, nerves, and lymphatics run in the submucosa, and the arrangement of the epithelium and submucosa varies with the tissue. Many folds and extensions are present in the small bowel, while the esophagus is relatively smooth.

The tight junctions normally prevent access of cells and proteins across the epithelium. Given the appropriate stimulation of mediator release from mast cells, these tight junctions can become loose. Loosening the junctions allows passage of phagocytic cells to help in defense. At these times proteins can also pass the epithelium, and this is the origin of the "leaky gut." The leaky gut is the explanation for the observation that at times people are more sensitive to some foods than at other times. It is presumed that more proteins cross the gut wall because of the loosened junctions between epithelial cells. This leads to more allergic symptoms and increases the chance of developing allergies. Direct damage to the gut wall by disease organisms, toxins, and autoimmunity will also lead to the gut being leakier.

Leaky Gut

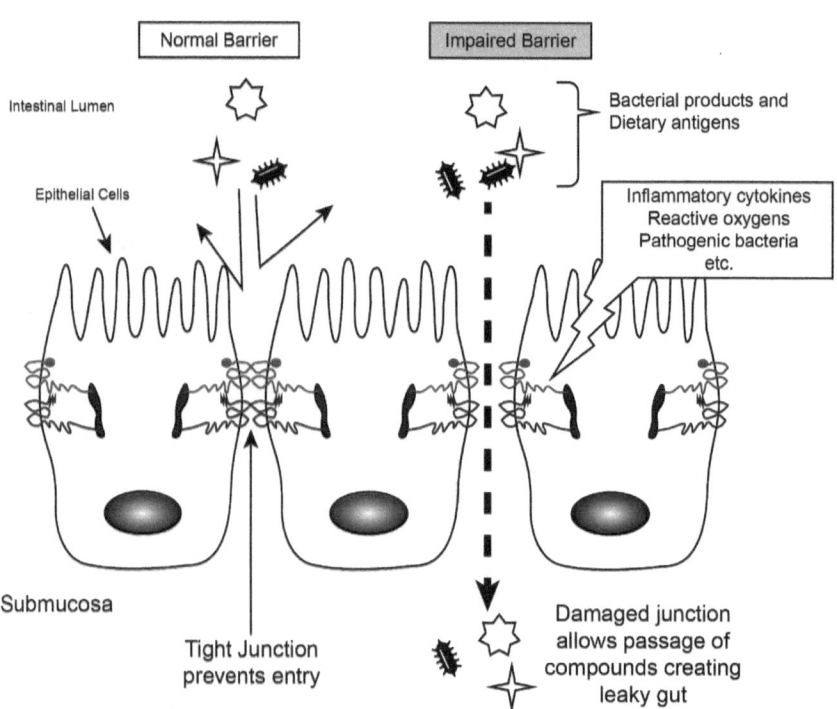

More about Antibodies

Antibodies are produced by B lymphocytes away from the site of where they will be used for defense. B lymphocytes are concentrated in tissues known as lymph glands. Lymph glands are found connected in rows, like beads on a string, and they act to drain fluid away from our body tissues. Lymph glands are part of the lymphatic system that cleans and monitors fluids that have leaked from our blood vessels into our tissues. By filtering the fluids before they are returned to the blood system, the lymphocytes detect what is happening in the body.

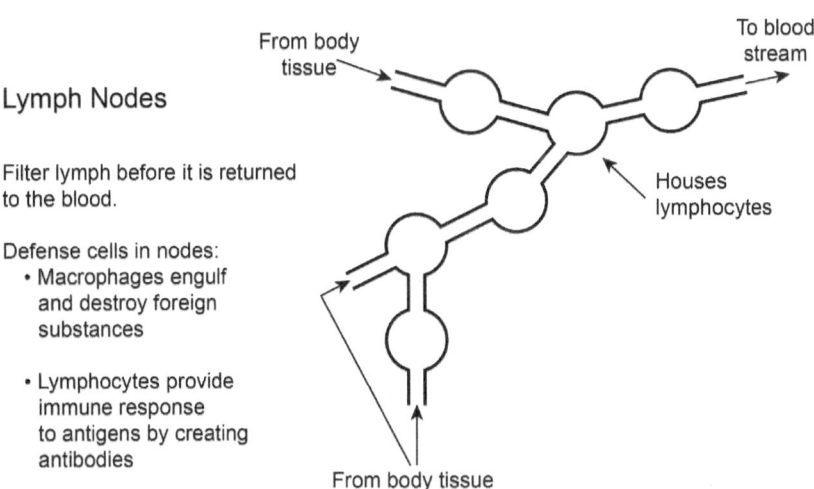

Lymph Nodes

Filter lymph before it is returned to the blood.

Defense cells in nodes:
 • Macrophages engulf and destroy foreign substances

 • Lymphocytes provide immune response to antigens by creating antibodies

From body tissue

To blood stream

Houses lymphocytes

From body tissue

The B lymphocytes reside in lymph glands and when stimulated respond by multiplying in number, enlarging the lymph gland. We might experience "swollen glands" at these times. Antibodies are created by the B lymphocytes, and the antibodies are transported from the lymph glands into the bloodstream, where they are distributed to tissues throughout the body. Antibodies have a specific immune response to specific antigens (foreign proteins that in the case of allergy are called allergens). Because antibodies are distributed throughout the body, skin allergy testing or blood allergy tests are useful and reliable indicators of the body's antibody status. The whole body is primed with antibodies, ready to respond to an antigen not just a specific region. It is the presence of the antigen contacting the body's antibodies that gives us the immune response, be it the nose with hay fever, the lungs with asthma, or the gut with food allergies.

Antibody Types and Functions

	IgM	IgG	IgA	IgE
% of total antibody in blood	6%	80%	13%	0.002%
Fixes complement	Yes	Yes	No	No
Function	Primary response, fixes complement.	Main blood antibody neutralises toxins, activates component system.	Sectreted into mucus, tears, saliva.	Antibody of allergy and anti-parasitic activity. Acts as alarm bell.

IgA is primarily secreted across the gut wall (mucous membranes) where it sticks to threats, preventing them from adhering to and entering the body. IgG defends within the body by specifically attaching to identified threats. Once identified and coated by the IgG, the threat is then engulfed by macrophages, neutrophils, and eosinophils, or the threat is eliminated by the complement system. The culmination of all these processes we generally consider to be inflammation. When these barriers are overwhelmed, the body is down to the final line of defense. In such an event, what would you do? Would you consider ringing an alarm bell, or would you send out a distress call? This is exactly what the body does. Our system uses the antibody called IgE. IgE is our final defense and acts like an alarm bell warning of a problem and encouraging other cells to come to the area to help with defense.

Dr. Neil B. Benson, MD

Process of Antibody Creation and Effect

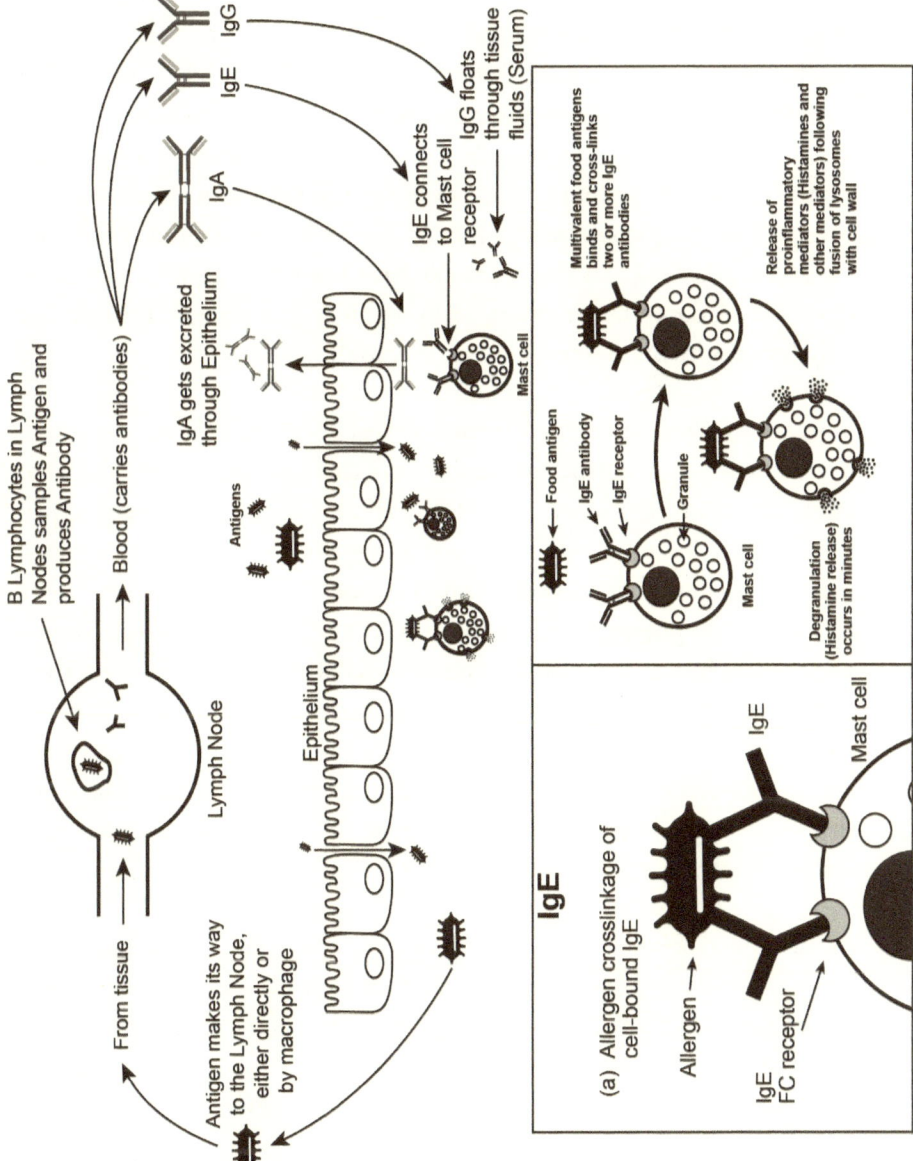

On the Origin of Diseases   25

IgE is mostly attached to mast cells (tissue macrophages) and these cells sit in connective tissues, usually just under the surface of the gut mucosa (submucosa) or skin. If these antibodies contact something they have been programmed to react to, the result is degranulation of the mast cell. Degranulation means the release of chemicals called mediators. These mediators include histamine (which is why antihistamines are used in allergic reactions) and many others that send out a generalized alarm message to all other defense cells to come here, as there is a perceived urgent threat. This results in the swelling and itching we experience in allergy. The allergic response is the final layer of defense and is initiated when the other barriers of protection have been breached. All antibody types are produced by lymphocytes in response to a perceived threat, but only IgE resides on mast cells (tissue macrophages).

Once mast cells are stimulated, mediators are released into the surroundings. These mediators tell other protective cells to come and defend the area. Protective cells are always present in the blood, and they are pumped through our blood vessels to all our body tissues. With IgE mediated mast cell degranulation several things happen. The blood vessels dilate and allow the phagocytic cells to come out of the blood vessels and to concentrate in the area. Phagocytes are also stimulated to increase their activity and effectiveness. Many different mediators, including histamine, are involved in the process. It is interesting to note that both parasites, especially worms, and the allergic reaction involve IgE. It just so happens that to develop allergies the body must first be invaded by a parasite. This is the connection; the parasite comes before the allergy! More on this later.

As mentioned previously, lymphocytes are typically found in accumulations known as lymph nodes. In the mouth and throat we also have larger accumulations of protective lymphatic cells in specific structures known as the adenoids and tonsils. In the small bowel similar large accumulations are known as Peyer's patches. These lymphocytes accumulations, large and small, make up part of the mucosa-associated lymphatic tissue (MALT). The MALT cells are constantly sampling and formulating a response to potential threats. When the body detects a significant invader, it may enlarge these lymph glands, and a sore, swollen throat may result.

Lymphoid Organs

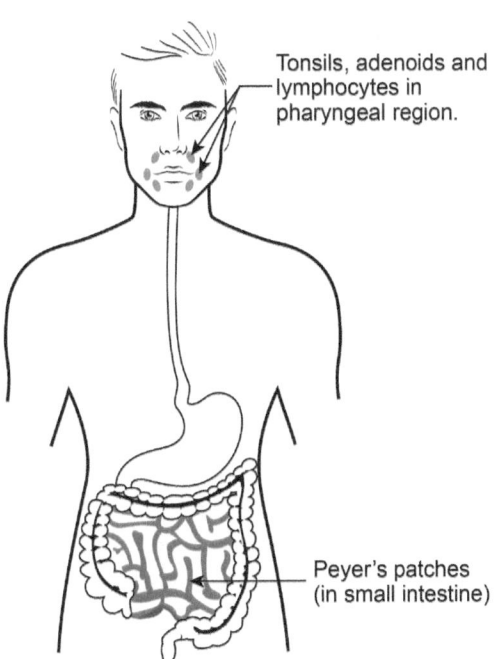

Mucosa-Associated Lymphatic Tissue (MALT)

• Includes:
  • Peyer's patches
  • Tonsils
  • Other small accumulations of lymphoid tissue

• Acts as a sentinal to protect respiratory and digestive tracts

Peyer's Patches "Tonsils of the intestines"

• Found in the wall of the small intestine
• Capture and destroy bacteria in the small intestine

Tonsils, adenoids and lymphocytes in pharyngeal region.

Peyer's patches (in small intestine)

Lymph gland enlargement might be diagnosed as tonsillitis, pharyngitis, while in the small bowel stimulation of the Peyer's patches can lead to large swollen lymph glands that can be mistaken as appendicitis (known as mesenteric adenitis). These lymphocytes, with the help of phagocytes, sample the environment and respond by producing specialized antibodies appropriate to the body's defensive needs. It has been estimated that 80 percent of the body's immune cells are located in the MALT system of the gut. It is hardly surprising that most MALT activity is around the small bowel (in Peyer's patches), an area responsible for the absorption of nutrients. By controlling biological competition in the small bowel, we improve our nutrient absorption and health. As so much of our immune function is directed at the gut (80 percent) we should keep in mind the gut when we talk about altered immunity. Hippocrates was right in his observation that "all disease begins in the gut," and now we are learning why.

Host: Pathogen/Parasite Models

We will now discuss what happens when a pathogen tries to obtain a living from our body (parasitizes us). The outcome is that either our body succeeds in defending against the pathogen, and it is removed, or the pathogen overcomes our defenses and either continues to live within us, or it kills us. It is rarely in the interest of a pathogen to kill the host quickly. Pathogens survive and thrive better if they keep the host alive at least for a time. The longer the host is alive the longer the pathogen continues to live and breed. This is the evolutionary outcome of most host/pathogen interactions and is classified into a variety of relationships. The interaction may be pathogenic/parasitic and of singular benefit to only the pathogen, or the relationship can be viewed as being symbiotic, also of benefit to the host. Whether pathogenic or symbiotic, the outcome of the host/pathogen interaction is colonization.

When the outcome is colonization, what is not taken into consideration by most doctors and scientists is the long-term effects of these nonlethal interactions, be they "parasitic or symbiotic." What I am proposing is a radical new way of looking at disease based on long-term interactions of the pathogen/host paradigm. In this model, what is important is how the pathogen interacts with the host over the long term. I have previously presented this model as the "Abnormal Bowel Cascade" (Benson, 1995).

The abnormal bowel cascade identifies how defenses are sequentially brought into play and how this affects the host over time. What is clear is that for allergic diseases and autoimmune diseases to develop, the host's general defenses must first be breached. Now the body recruits additional lines of defense involving cellular defenses and lymphocytes. Lymphocytes can be directly involved in the defense either as T cells or B cells. In the case of B cells, three main types of antibodies are produced, and they are called, IgA, IgG, and IgE. This model recognizes that one pathogen can result in many different diseases, depending on what immune defenses are triggered.

Different immune systems (based on genetics) put up different preferences for defenses. The same person's immune system also acts differently depending on the length of time of the parasitic interaction. If the infection is not cleared by one immune response, a further immune response will be created until, ultimately, IgE is used. This understanding gives a model for studying allergic and autoimmune diseases, and, more importantly, gives us an approach that can be used to heal the resultant diseases.

To move to a causal model for allergies and autoimmune disease, we must look at the immune system and the relationship to pathogens. The immune response is there to defend us against a perceived threat. The allergic or autoimmune inflammatory response is the result of our body trying to defend itself. Ultimately, it is a pathogen that should be treated in inflammatory illness presentations. This is in opposition to traditional treatment, which aims to dampen down the immune response to remove the symptoms. Addressing the pathogen moves from symptomatic (non-curative) to causal (curative) treatments.

With this causal way of thinking, the study of allergy largely disappears, as do many of the disciplines that traditionally make up medicine. We move from treating symptoms to treating causes. Move from medications that block the immune system (such as antihistamines, steroids, anti-inflammatories, and disease modifying drugs such as methotrexate), to cure by addressing the underlying causative agent.

While there is considerable benefit from treating symptoms, I believe that many people, when given an option of cure, would choose this approach over only disease management—the symptomatic approach. People will ask the question, "What pathogen is causing this disease?" rather than viewing immune modulation of the symptoms as the only treatment option.

Examples of Chronic Infections that Cause Disease

Medical doctors have long observed how diseases can change over time. Syphilis, which was known as "the great pretender" or "the great imitator," has numerous presentations depending on an individual's immune system and the length of time he has had the disease (called first, second, latent, and tertiary stages of the disease). Initially symptoms present as a painless sore known as a chancre, which disappears after a few weeks. The second stage may overlap with the presence of the chancre, or it may start weeks later. Presenting as a rash, typically on the hands and bottom of the feet, the second phase may have an array of additional symptoms including a sore throat and fever, swollen lymph glands, fatigue, muscle aches, and other signs including skin lesions, called condyloma lata, patchy hair loss, and many more. A latent phase typically follows in which there is no sign of the disease, and this may last from months to years. The final or tertiary phase occurs years after the infection is acquired and can affect a variety of organs including the brain, nerves, eyes, blood vessels, heart, joints, bones, and liver. The ultimate presentation depends on the organ that has been affected. Syphilis has a massive array of presentations and lives up to the name "the great pretender."

What I have come to understand is that many other pathogens also have chronic stages of infection and thus different symptoms presenting with time. Traditionally we have named different diseases based on symptom groupings and on the end organ that is involved, be that skin, bowel, eyes, lungs, etc. With such an approach we have failed to note the common causation of many different diseases. A system based on disease symptoms is incomplete and often inappropriate for studying and researching chronic disease, especially chronic inflammatory disease. We must move to a system that looks at causation.

An understanding of the chronic nature of infection appears to underpin an understanding of all autoimmune diseases and allergies. For a pathogen to breach the defenses of gastric acidity and small bowel enzymatic breakdown, it will have special attributes. It may hide in the gastric mucosa under the mucous as in the case of *Helicobactor pylori*, or it may have a special coating like tape worms or change its surface antigens like malaria, or it may live intracellularly like chlamydia or rickettsia. Others, such as C. albicans, have a yeast phase, which lives on the surface of the gut essentially out of reach of much of our immune system's weaponry. All human pathogens have some tricks that allow their continued survival at the expense of at least some of the members of our species.

H. pylori is one such organism, and a book could be written on this topic alone. Professor Barry Marshall, a physician who along with a pathologist, Dr. Robin Warren, both from Australia, defined the role of H. pylori as a causative agent in stomach ulcers. Both doctors were awarded a Nobel Prize for their valuable contribution to medicine. Professor Marshall goes on to say, "When the work was presented, my results were disputed and disbelieved, not on the basis of science but because they simply could not be true. It was often said that no one was able to replicate my results. This was untrue but became part of the folklore of the period. I was told that the bacteria were either contaminants or harmless commensals." (Marshall 2005). As we can see in this case, medicine reluctantly followed scientific research.

Sometimes, more than one chronic infection can be involved in some diseases. Dual infections can be found with the story of the Burkitt's lymphoma. Dr. Burkett noted that the Burkitt lymphoma in Africa is likely the result of a malarial infection and the Epstein Barr virus (Kellock 1985). Dr. Burkitt noted that a very quick-growing cancer on the faces of children in parts of Africa, known as Burkitt lymphoma, was found only in areas where malaria occurred. He also noted that in every cancer he checked, the

Ebstein-Barr virus was found. If only the Ebstein-Barr virus was present, a child would not get the lymphoma. For the lymphoma tumor to occur, the malarial infection must also be present. This tumor has the need for two infections; otherwise it did not develop. The tumor also occurred only in children, meaning it was also age dependent.

Dr. Denis Burkitt, a distinguished British surgeon, was told that "there was a general reluctance amongst editors of medical journals in the early 1970s to publish any article which attempted to link by a common cause those diseases which medical opinion generally regarded as quite separate. Denis was told categorically that no medical journal would be willing to accept any article linking a number or totally different diseases by a common cause" (Kellock 1985, 149).

With the warning from the medical journals to Dr. Burkitt in mind, it can also be pointed out that in some cases the same symptom complex, or disease, might have several different causative organisms. Rheumatoid arthritis is such an example. Rheumatoid arthritis is a disease that can be caused by several different organisms, including C. albicans and the bacteria Klebsiella pneumonia.

Some diseases are the result of chronic infections and the presence of proteins that have crossed the gut barrier. This appears to be the basis of type one diabetes. An environmental factor (most likely a parasitic infection, and I suspect C. albicans) and the milk protein casein are both necessary for the creation of the disease. If you are interested, it is fascinating to explore the A1 and A2 milk controversy, and it will be discussed later.

Finally, depending on how the immune system reacts through time, a given disease may disappear only for another to develop. Such is the case with childhood eczema, which often disappears only for the patient to develop another disorder, often asthma. The underlying causal pathogen has not gone; rather, the immune response has changed, and the symptoms are now affecting a different organ, creating a new disease. I have no doubt this is the basis of rheumatoid arthritis when over time the disease changes and "burns itself out." The disease has changed because the immune reaction has altered, while the underlying causal organism may very well still be there. It is an inconvenient truth, but pathogens underlie most if not all allergic and autoimmune disease.

For a pathogen to cause an allergy or autoimmune disease, the pathogen must have the potential to chronically parasitize the host. Pathogens with a short-term presence in the body are unlikely to lead to chronic disease. With an acute infection, in some cases an organ can be irreversibly damaged. The damage created by an organism or our immune system permanently and irreversibly changes an organ, and permanent disease results (e.g., heart disease, caused by Q fever). Once the infection is eliminated the structural damage persists. These structural changes are, however, no longer inflammatory. All allergic and autoimmune diseases are inflammatory. They all have an underlying pathogen which is still present providing an ongoing inflammatory immune stimulus in active disease. The chronic presence of the pathogen leads to the inflammatory disease. These diseases can change over time depending on how the immune system responds and which body organ is affected.

Some Pathogenic Organisms that Can Stimulate Chronic Immune Reactions

This table of organisms that can cause chronic diseases is primarily meant for scientists. It involves fact and conjecture, but as they all have the capacity for chronic infection, there will be a chronic disease association. Some are known now, as indicated, and others are yet to be discovered. This table is meant as food for thought and is hopefully perceived as fertilizer to stimulate scientific discovery rather than seen as feces. Enjoy!

Listing of Chronic Infective Organisms Suspected of Chronic Diseases		
Family	**Species**	**Disease**
Spirochetes (The Great Imitators)	*Treponema pallidum* (Syphilis)	Many diseases including Alzheimer's dementia
	Borrelia burgdorferi (Lyme's disease)	
Rickettsia's (Intracellular organisms)	*Rickettsia prowazekii* (typhus)	Unknown
	Coxiella burnetii (Q fever)	
	Rocky Mountain spotted fever	
	Chlamydia	
	Other unknown pathogens	

Fungi	*Candida albicans*	Rheumatoid arthritis Crohn's disease Multiple sclerosis Parkinson's disease Alzheimer's disease Endometriosis Allergies (plus high uric acid) Gout Many others
	Other Candida species	Unknown
	Aspergillus	Unknown
Bacteria	*Salmonella typhi* (chronic diarrhea)	Ankylosing spondylitis Psoriatic arthritis Other seronegative Spondyloarthropathies Reactive arthritis Reiter's syndrome Guillain-Barre syndrome
	Klebsiella pneumonia	Rheumatoid arthritis Ankylosing spondylitis Crohn's disease Ulcerative colitis Implicated in heart disease
	Helicobacter pylori	Stomach ulcers Gastric adenoma Lymphoma
	Neisseria gonorrhoeae (gonorrhea)	Pelvic inflammatory disorder Infertility

Other		
Other	Mycobacterium paratuberculosis (avium)	Crohn's disease Ulcerative colitis
	Parasitic worms (tapeworms, hookworms, and other helminth infections)	Allergies
Protozoa	Malaria (plasmodium sp.) Malaria plus Epstein Barr virus	Unknown Burkitt lymphoma (some)
	Toxoplasmosis	Unknown
	Cryptosporidium	
	Giardia	
	Trichomonas vaginalis	Prostate cancer
Viruses	Epstein Barr virus	Various cancers
	Hep B and C	
	Human papillomavirus	
	Many others	
	Ross River virus	Unknown
	Barmah Forest virus	
	Others	
	HIV plus other pathogens	HIV

Dr. Neil B. Benson, MD

Chronic Disease

Infective agents causing chronic disease have special mechanisms for evading destruction and colonizing the host. Successful colonization also depends on the immunologic health of the host. Many organisms that were not previously considered as human pathogens become recognized as pathogens in the immunocompromised. Pneumocystis carina and many other "bugs" in people are now recognized as pathogens in patients with HIV/AIDS. As we are dealing with chronic disease of the otherwise healthy, not the immunocompromised, we will therefore ignore this interesting area. It is this chronic colonization of the host by the pathogen and the interaction with our immune reaction that results in what we know as inflammation, allergies, and autoimmune disease.

Candida Albicans, the Stealth Pathogen

Candida albicans (C. albicans) is a fungal organism, and the name literally means white white, (Candida in Latin and then albicans in Greek). C. albicans is a major pathogen in the Western world today, and it is often referred to as candida or even thrush. C. albicans is not recognized by mainstream medicine today as a major contributor to the burden of poor health. Some alternative health practitioners have, however, long recognized C. albicans as a causal agent in human disease. It was through this alternative pathway that I first heard of C. albicans causing chronic disease. Initially I was skeptical of the claims that C. albicans was involved as a significant pathogenic agent, as C. albicans was largely ignored in my medical training. Skin irritation, diaper (or nappy) rash, and vaginal thrush were recognized as inconveniences. I was also taught that in immune-compromised individuals, C. albicans could cause a serious disseminated disease, leading to death. Between these two extremes the information available about C. albicans as a chronic infection able to cause disease was at best considered fringe. Information available to me was poorly formulated and not scientifically accurate. The lack of accurate scientific information allowed me and other doctors to dismiss the information on "chronic candida." C. albicans was therefore not recognized as a causal agent of chronic disease. I can understand how many colleagues still hold this view, as I too, early in my career, dismissed patients that said they were affected by "chronic candida." I should have known better, as I had extensively studied fungal organisms in my university education, in botany and plant pathology. I had seen that in the plant world fungal organisms are a major disease burden with a huge variety of presentations.

It was not until I was presented with Dr. Crook's book *The Yeast Connection* that my opinion changed on C. albicans. Dr. Crook is a medical doctor from Jackson, Tennessee, in the United States of America. He put science into the Candida debate. His series of books known as the "Yeast Connection" included books with titles such as *The Yeast Connection and the Woman*, *Chronic Fatigue Syndrome and The Yeast Connection*, and others. Many were on the bestseller lists for a very long time. The information Dr. Crook presented was very readable and convincing. While his books might not have been accepted by the traditional medical practitioner, he did put an indelible mark on my formative post-medical-school education. It is ironic that a nonscientific publication written for the public should have played such a role in my journey for the cause of allergy. Thank you, Dr. Crook!

I can understand Dr. Crook's approach of presenting information directly to the public, rather than publishing it in journals. He no doubt sought to avoid the pitfall that Dr. Burkitt encountered where the scientific community's 'one-cause many-disease' thinking is not allowed. Many other people have also made valuable contributions to this difficult area of study. Dr. C. Orion Truss, a specialist in internal medicine from Birmingham, Alabama, USA, is one of many who should be recognized for his role as a pioneer in the field of candida research.

From these pioneering observational medical practitioners, we will now move on to a systematized understanding that can be tested. It is my hope that the information that I am providing will move C. albicans out of the alternative health fringes squarely into the mainstream medical arena. C. albicans can then finally be recognized as a major public health issue that it is.

To understand how C. albicans is a disease-causing organism you must think ecologically. Ecological thinking is all too often lacking in traditionally trained medical practitioners. If you run into a doctor that is dismissive of your inquiries about C. albicans, please persevere and show them my work. The book is written in a style that hopefully is both interesting to the public and scientifically rigorous enough that your doctor can learn and increase his or her understanding. It is important to have a doctor that is open-minded, willing to listen to you and learn about your health issues.

The Ecology of Candida Albicans

C. albicans cannot exist outside of the body of an animal for long. It requires an animal host for survival and proliferation. While it is hard to prove, estimates suggest that C. albicans exists in about 60 percent of the human population. C. albicans is a normal inhabitant of the human gut, which can at times act as a pathogen. C. albicans may be present in small amounts where it is controlled by the body defense mechanisms and by the gut flora. When controlled, C. albicans is a commensal, and it is not considered damaging to the host. At the other end of this scale is invasive C. albicans, which can cause death. What we are discussing is the process by which a benign commensal can be activated to become a killer. This process is common and leads to a spectrum of diseases that are best described by what we earlier referred to as the abnormal bowel cascade.

The abnormal bowel cascade is the systematic loss of barriers of defense leading to immune stimulation, vitamin deficiencies and allergies and autoimmunity. This shows how the defense mechanisms of the body can be overcome or selectively impaired by various means. Antibiotics directly destroy the gut flora, poor diet weakens the gut flora, and malnutrition reduces body resistance. Genetic and acquired immunodeficiencies further lower defenses. With impaired host defenses, C. albicans has an easier time surviving, and it is then able to increase in number. As the body is increasingly colonized by C. albicans, our body brings in more defensive strategies to counter the C. albicans infection. It is this process of recruiting further immune mechanisms that ultimately leads to the development of allergies and autoimmune disease. There will be more discussion on this later.

How Is Candida Albicans Stealthy?

Unlike most pathogens, C. albicans is a fungus and thus a eukaryotic organism. In other words, it is much more closely related to us genetically than viruses and bacteria. Viruses and bacteria are what we usually think of as being human pathogens. As a eukaryote, C. albicans has a cell nucleus and many other parts that give C. albicans a similarity to our own cells. This similarity means that there are fewer points of difference with their structure as compared to viruses and bacteria. Our immune system is trained to look for points of difference. Difference allows our body's immune systems to recognize a foe. It is for this reason that many bacteria and viruses can be contained and eliminated by the body with relative ease. The more points of difference you have available, the more immune strategies you can develop to remove or thwart a pathogen. The more similar something is to us, the fewer opportunities and strategies we have available to us to fight off the intruder. By being a eukaryote, C. albicans gains its initial stealthy character. It is so like our own cells that our body largely views C. albicans as a friend.

As discussed earlier, it is usually not in the best interest of a parasite or pathogen to quickly kill its host. C. albicans is no exception. It is in the best interest of C. albicans to appear benign, hidden from the radar, able to avoid setting off the defense mechanisms of the host.

Stealth is gained not only by C. albicans being a eukaryote but also by it having the advantage of different forms of growth. An active yeast form lives on the gut mucosa and skin surface while a mycelial or root-like form can invade between human cells and may even progress to forming a mat. There is also a dormant form that produces outgrowths that interlock forming resistant protected balls of candida. The adaptive nature of C. albicans makes it a worthy opponent. The variety of forms allows C. albicans to live benignly on the host as a yeast, or it can invade the host as a mycelium should the conditions allow, and finally it can lay dormant waiting for conditions to improve. We clearly have an opportunistic competitor.

**Candida albicans Mycilial Hyphae Invading Gut Wall
Triggering Mast Cell Degranulation**

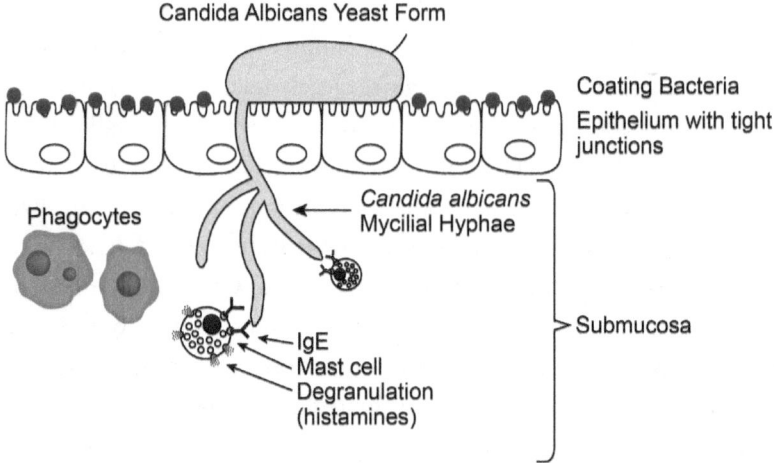

Candida Albicans Yeast Form

Coating Bacteria

Epithelium with tight junctions

Phagocytes

Candida albicans Mycilial Hyphae

Submucosa

IgE
Mast cell
Degranulation
(histamines)

Dr. Neil B. Benson, MD

So, How Do We Get Infected by C. Albicans?

The most common time that C. albicans is acquired is at birth when contact with the maternal vaginal microflora occurs. To explain how C. albicans is acquired at birth we must first understand that the vagina of healthy mothers is not sterile but is full of bacteria and other biological organisms. Most of the organisms in the vagina of fertile women are bacteria from the species Lactobacillus acidophilus. A symbiotic relationship exists between L. acidophilus and women. The vagina mucosa excretes sugars that are used by the L. acidophilus to make lactic acid, which acidifies the vagina. Normally the pH goes down to 3.8 by this process. The name Lactobacillus acidophilus literally means milk-fermenting bacteria and acid loving. In an acidic medium C. albicans and most human pathogens find it hard to grow and are outcompeted. A recent article by O'Hanlon et al (2019) in *BMC Microbiology* says it quite nicely when it states: "The composition of the vaginal microbiota is known to alter dramatically a women's resistance or susceptibility to reproductive tract infections. Women with a low Nugent score (predominantly lactobacilli morphotype) microbiota are at a reduced risk of most sexually transmitted infections including HIV-1, gonorrhea, and trichomonas, as well as obstetric infections that contribute to preterm deliveries and perinatal complications, infection by bacteria implicated in pelvic inflammatory disease and other gynecologic infections and urinary tract infections indicating that the vaginal lactobacilli have a broad antimicrobial activity against viral, bacterial and eukaryotic pathogens in the reproductive tract." Clearly a vagina colonized with L. acidophilus offers significant health benefits not only to the woman but as we shall see also to her baby.

While living in the mother's uterus, the baby is kept protected and the environment is normally sterile and free of bacteria. The surface and gut of a baby is devoid of bacteria. With a vaginal birth and transit down the vaginal tract, a baby picks up the organisms from the mother's vagina. These organisms coat the surface of the baby and enter the oral cavity, beginning the process of gut flora establishment. A mother having C. albicans, or a poor vaginal microflora, is an obvious problem. Not being colonized by beneficial bacteria delays the establishment of the protective gut flora and opens the opportunities for organisms to establish that are less beneficial. The statement "that nature abhors a vacuum" is true; in other words, if there is a space for something to live, then something will live there. If beneficial organisms are not occupying the space, then something else will fill it up. If C. albicans is present in the vagina, then C. albicans will become established on and in the newborn at this time.

For arranged C-sections, there is no contact with maternal vaginal organisms. The first coating of bacteria will be from whatever the baby contacts. C-sections are a recognized risk factor in the development of allergies in babies. The lack of contact with beneficial maternal vaginal bacteria and subsequent colonization by less beneficial organisms gives a mechanism to explain this observation.

Emergency C-sections, unlike arranged C-sections, will usually have some limited contact with the maternal vaginal flora. With rupture of amniotic sac membranes, vaginal organisms contact the baby, and at least partial colonization occurs.

The area of perinatal exposure to vaginal flora is an active field of research. In the future it will be beneficial for all babies to routinely be exposed to beneficial strains of bacteria at birth. This will properly establish a baby's gut and skin flora. At present there is a program for checking for strep B a known vaginal pathogen that causes infections in newborns. This program should be extended to include C. albicans as a part of prenatal maternal health checks. Reduced exposure to C. albicans at birth will reduce the establishment of C. albicans in babies and will benefit the baby's long-term health.

Once present in the gut of a baby, C. albicans will compete with the newly establishing gut flora. If the gut flora is strong the C. albicans will be inhibited and no long-term problems may develop. With a weak gut flora, C. albicans will establish and create immune damage. The ultimate outcome will depend on several factors in addition to the presence or absence of beneficial gut organisms. Breastfeeding has many benefits to the baby beyond just providing nutrition. The baby obtains antibodies not only in the colostrum but throughout the time of breastfeeding. Colostrum is the first milk produced by the mother. Colostrum looks watery compared to later breast milk, and it is filled with maternal antibodies. Many antibodies are directly absorbed by the intestinal tract of the newborn, taking up residence within the baby's immune system. Having maternal antibodies provides defense for the baby until such a time as the baby develops its own immunity. As the baby ages, the number of antibodies absorbed from the breast milk into the baby declines. While not absorbed, antibodies in the breast milk still help defend the baby from infections in the gut. Breast milk also has compounds that are not absorbed by the baby, and these compounds act as prebiotics, actively feeding and promoting the gut flora.

If the vagina becomes colonized by C. albicans or other species that are pathogenic or potentially pathogenic, such as strep B, a simple trick can be used to aid in the reestablishment of the normal vaginal flora of L. acidophilus. Place L. acidophilus yogurt on a tampon and insert a new yogurt-coated tampon nightly for three nights. In my experience this technique coupled with appropriate medications works to help restore the L. acidophilus and displace the pathogens in affected patients.

C. Albicans as a Pathogen

C. albicans has been present with mankind for a very long time. It is only recently that this organism has had the opportunity to express its full potential as a pathogen. Antibacterial antibiotics and poor diet reduce the beneficial gut flora and allow for overgrowth and establishment of resistant organisms. C. albicans is one organism that is resistant to what most people think of as antibiotics. With less competition being exerted by a damaged gut flora, C. albicans overgrows, establishes, and invades the host. This invasion causes the recruitment of IgE and other immune defenses, and this is the likely causal explanation for the rise in allergy and autoimmune diseases. The mechanism is there for the rise in allergies and autoimmune disease, and we can turn back the clock on these diseases if we take C. albicans into account. Going forward we should be preventing allergy by preventing damage to the gut flora, and we should remove C. albicans in children at an early stage. It is a mistake to assume that C. albicans is a normal component of the gut flora and not worthy of eradication. From a public health point of view, cure is better than symptomatic treatment, but prevention is better than cure. Why take antihistamines for hay fever when we could get rid of the allergy or, even better, prevent it from ever developing.

Many theories have been put forward to explain the rising rates of allergic disease, including the "excessive cleanliness" theory. This theory has been debunked, and I want to stress that it is the overuse of antibiotics and poor diets that primarily account for the rising rates of allergies and autoimmune disease. The damage to the gut microbiome, the resultant increase in gut pathogens, and the stimulation of the immune system via the process of the abnormal bowel cascade are responsible for most of the rise in allergies and autoimmune diseases.

Another controversial area that C. albicans is involved in is cancer. This is not my area of expertise, but it makes sense that if you have a pathogen compromising your immunity then getting rid of it will improve your chances of fighting off the cancer. It may be that C. albicans is directly causal in some cancers, but I do not have evidence to support this. Viruses are certainly implicated in the cause of many cancers, and it might be that C. albicans sufficiently compromises some people's immunity so that the cancer can become established. Some alternative cancer clinics treat C. albicans routinely, at the start of cancer treatment. C. albicans is also often treated during traditional cancer treatments. Either way the health of individuals improves with getting rid of C. albicans.

Chronic bacterial infections are implicated in prostate cancer and chronic Helicobacter pylori infections in some stomach cancers. Both cause chronic inflammation and cell death, and this is increasingly being recognized as a cause of cancer. Might not chronic C. albicans infections also be a source of chronic inflammation leading to cell death and cancer?

C. albicans also produces compounds that either act directly as toxins on our body or indirectly via our body's immune system. How endometriosis is caused by C. albicans is not well understood by me, but I assume it either produces hormones or compounds that block or stimulate receptors for the hormones in the body. Endometriosis and C. albicans often go hand in hand, and I have found that the removal of C. albicans stops the development of endometriosis. Clearly to me a causal relationship is present. Many cases of female and male infertility are also affected by C. albicans. Treating the C. albicans will often result in improved fertility.

While antibiotics have a very important role to play in human medicine (and there are many different types: antibacterial, antiviral, antiprotozoal, antifungal, to name the main groups), many medical doctors have in my experience been too liberal with the use of antibacterial antibiotics (penicillin being a name you are probably familiar with), and have generally failed to recognize the importance of antibiotics on the gut flora. Clearly if antibiotics have damaged the gut flora, you have impaired an important defense mechanism for the body and have left the taker of the antibiotic potentially vulnerable to gut infection. The effect of poor diet also has a negative effect on the gut flora. Poor food choices are closely correlated with increases in almost all diseases. By looking at the effect on the gut flora we can see how a poor diet can negatively affect our own health.

Chemical sprays on food is an area in need of further study. Sprays are applied in the field, by the agricultural community, with the aim of preventing plant or animal diseases. Chemicals are also applied with the aim of preventing food spoilage. The effect of most of these substances, which we generally call pesticides, herbicides, and preservatives, has not been adequately studied to see their effect on the gut flora. This effect needs to be thoroughly studied.

Our drinking water likely needs review as well. The effects of chemical additives on the gut flora is a complex area. While I am not against low levels of chlorine in water as a public health measure, I am not sure that the impact of high levels on the gut flora have been studied or that a safe level has been determined. The sole purpose of chlorine in drinking water is to kill bacteria, algae, and other organisms. Some children accidentally or intentionally drink significant amounts of water when swimming. High concentrations of chlorine, especially found in pools, may have some effect on the gut flora and the skin flora. This area needs further study, and if chlorination is found to be damaging to the gut flora then we should consider alternative water treatment options. Options could include reverse osmosis or filtration, at least for our drinking water.

Treatment of Allergies

For treatment of allergies and autoimmune disease to be successful, three factors must be addressed simultaneously. Successful treatment involves eating a healthy diet, allergen avoidance, and the appropriate use of medications. Good diet is essential for establishing the beneficial gut flora and is always the starting point.

We can also improve the gut flora by reseeding the gut with probiotics or by consuming products that contain beneficial bacteria. Sauerkraut and kimchi are both made from fermented cabbage. Bacteria used in making these fermented products can help reseed the gut. Some of the bacteria can live in the gut and when eaten will become a part of the gut flora. Yogurts have also been used for their probiotic benefit. Yogurts made from bifidobacteria species can colonize the gut, as can lactobacilli species. While L. acidophilus has a major health role in the vagina, it has a lesser role in the gut. When eaten, lactobacilli and bifidobacteria have beneficial, anti-inflammatory properties in the gut.

The use of probiotics and other reseeding strategies such as crapsils (fecal transplant capsules) and fecal enemas have scientific merit. The intent is to recolonize the lower bowel (colon). Keep in mind that once C. albicans is established in and on the gut surface you can take all the crapsils you want, but you will never get rid of C. albicans. C. albicans can live from the tip of the tongue all through the bowel.

Several different microbiomes exist throughout our body. The lungs have their own protective microbiome as does the prostate. Several others exist throughout the gut. Special organisms live on the teeth, while others are found in the mouth, nasopharynx, esophagus, small and large bowel. We have already talked about the importance of the large bowel microflora, and it is worth remembering that other microbiomes also exist.

While gut floras inhibit C. albicans, they are not always able to eliminate C. albicans. When C. albicans has become invasive through the epithelial (gut) wall, the gut is no longer in direct contact, and competition with C. albicans does not directly take place. A colonized mouth, tongue, or esophagus is a significant problem. An infected tongue acts to seed the rest of the bowel whenever C. albicans organisms are swallowed. These organisms can then establish themselves lower down the gut should the appropriate conditions be available.

Many bowel flora tests are being marketed. Unfortunately, these are of limited benefit when it comes to identifying C. albicans. When C. albicans is invasive in the gut wall and doing damage, there may be no sign of the yeast form present in the feces. If C. albicans is in the esophagus it is unlikely that it will be spotted in the feces, as it would have had to survive the gauntlet of digestive enzymes in the small bowel. The mycelial growth form spreads locally and is difficult to dislodge, so it will also not turn up in gut samples even when it is there.

Dr. Neil B. Benson, MD

Treating Food Allergies

"Before you heal someone, ask him if he's willing
to give up the things that make him sick."
—Hippocrates

If you have a significant food allergy, you should avoid consuming that food. If you eat what you are allergic to, you waste immune resources fighting the food. You may or may not get obvious symptoms from consuming the allergen, but either way you are likely to damage the gut surface through inflammation caused by your immune response. A damaged gut surface leads to more places for C. albicans or other pathogens to establish and leads to poor nutrient absorption and can possibly lead to nutrient deficiencies. Nutrient deficiencies then further compound the issue by reducing your ability to fight the invader. It is a vicious cycle.

We must not only eat a good diet and avoid allergens, but in most cases, medications are also required. The medications depend on the pathogens present. C. albicans is frequently implicated and while anti-C. albicans medications can be prescribed by doctors, antifungal compounds are also found naturally in a multitude of plants. Some of these plants can be used medicinally when they are eaten raw, cooked, or extracted. They may even be encapsulated or powdered and used by doctors and alternative health practitioners. I have tried many natural antifungals and will discuss how some are used as medications in a later chapter.

Treatment of Candida Albicans/C. Albicans

In all but the simplest cases C. albicans should be viewed as part of an ecological setting. We must consider all parts, which include gut flora, allergies, and medications.

It is beneficial to investigate why (for what reason) the person has developed problems with C. albicans. Has C. albicans been present since birth or was it acquired, was it a problem as a baby, and was it related to poor diet/ excessive use of antibiotics or both? Are nutrient deficiencies present? Is the patient a diabetic? When we have C. albicans, it is likely that we are going to have to make some changes to our diet and lifestyle in addition to the use of antifungal medications. The changes required will depend on our immune status, allergic burden, and nutrient status. I do not recommend taking antifungal medications without addressing the whole person, which includes allergies and diet.

In treatment some generalities will work for everyone. Eating a healthy diet is the starting point, and from there treatment must be individualized. I use the analogy of a mosaic. Putting together the patient's personal treatment program will have some pieces that are the same between people, but the way the pieces are placed is different from one patient to another. Your mosaic, or "picture of health," will ultimately have specific steps that you must take to return to health.

The background to all plans is the need to eat a healthy diet high in green vegetables. As mentioned earlier, the green vegetables promote the beneficial bacteria by feeding them soluble fiber. A healthy gut flora is key to regaining health and preventing recurrences. Food allergies compromise your immunity by wasting immune resources. Resource shortages make it harder to fight against C. albicans. If a food allergy is moderate or greater, and you do not avoid it, you will not get rid of C. albicans despite eating a healthy diet and using antifungal medications. Allergies not only waste our immune resources, but they reduce nutrient absorption. If nutrient deficiencies result, you will find it much harder to eliminate C. albicans. For the remainder of the book we will talk about allergies and then the medications used for treating C. albicans.

Allergies

So how did the term *allergy* come about? Allergy was first described in 1905 by an Austrian pediatrician, Clement Von Pirquet (1874–1929). Prior to Von Pirquet's work it had been recognized that something in the blood, referred to as antibodies, conferred immunity to infection. If a patient recovered from an infection and subsequently was reexposed to the same infection, he would not get the infection a second time and now had immunity. Von Pirquet studied this phenomenon in patients who had had TB. He found that two types of reactions occur when TB proteins were injected under the skin in patients previously exposed to TB. Skin reactions were either immediate (occurring in minutes) or delayed, occurring over forty-eight hours or longer after exposure. He termed both reactions as being "allergy." Allergy literally means "other activity" (allo meaning other, different, strange, and ergon meaning activity or reaction). Von Pirquet defined the allergic reactions as being of two types. The immediate reactions were of no apparent benefit to the host, while the late reaction conferred immunity to TB. We now refer to these reactions as being immediate antibody mediated and delayed cell mediated.

Von Pirquet would be disappointed by how his term *allergy* is used today. To outline this situation, I will refer you to the abstract of an article that appeared in *Allergy* titled "The history of the idea of allergy" (Igea 2013, 966-973).

> About 100 years ago, a young paediatrician understood that the function of the immune system should be rationalized not in terms of exemption of disease but in terms of change of reactivity. He coined a new word to

represent such an idea: "allergy": the first contact of the immune system with an antigen changes the reactivity of the individual; on the second and subsequent contacts this change (or allergy) can induce a spectrum of responses from protective (literally, immune) to hypersensitivity ones. The idea was at first hardly understood by the scientific community because it undermined the essentially protective nature of the immune response as it was defined. Nevertheless, in the next years, the growing clinical evidence led to the acceptance of this new point of view, but not in the new word, at least not unconditionally. The original significance of the neologism "allergy" became perverted and limited to describe hypersensitivity conditions. Perhaps because of the corruption of the term, today "allergy" does not have a well delimited significance among health professionals. Furthermore, the word has long ago escaped from the physicians to the streets, where it is popularly used also as synonymous with antipathy and rejection. This vulgarization of the term "allergy" has significantly increased its imprecision.

It was not until 1967, with the identification of IgE, that the term *allergy* formally became narrowed, referring only to IgE mediated immediate reactions. Thinking of allergy in terms of IgE alone is naïve, as it excludes many other immune-based reactions. Cell mediated and those related to IgA and IgG antibodies will be excluded from the now restricted definition of allergy. Labelling these non-IgE reactions as intolerance (as in gluten intolerance), has brought great confusion to the field. With the present lingo of allergy and intolerance, we understand that allergy is immune based, reliant on IgE, but we fail to appreciate that most intolerance is also immune based, just reliant on different antibodies (IgA and IgG) or T cells. Von Pirquet meant the term to include both protective and nonprotective hypersensitivity reactions. I think it is fair to say today that the term even in the scientific arena no longer includes protective reactions. "The widely accepted use of the word "allergy" today, where it is restricted to specific immunologic hypersensitivity reactions against harmless foreign antigens" (Huber 2006, 573-579). When talking of allergy, we will return more closely to the original broad definition of allergy. This broader definition includes all immune defenses directed against allergens.

Allergy therefore refers to all immune-based reactions that appear to have no benefit to the host. What benefit is there in having an allergy to grass, house dust mite, molds, or foods such as tomatoes and peanuts? I can see none. So why have we developed an immune reaction to these non-dangerous allergens? We become "allergic" to these things accidentally, as the result of a battle between our bodies and chronic infections. C. albicans is at the top of this list of the chronic infections causing allergies.

We can briefly summarize C. albicans once again. C. albicans is not needed for our survival. The relationship appears to be strictly one of host and pathogen/parasite. Though like my previous experience with the appendix, it may be that some yet undetermined value exists when C. albicans is in the controlled yeast state. Animals, including humans, pick up this pathogen from contact with another infected animal/human. C. albicans is what is known as an obligate parasite, and it has no ability to live outside of an animal host.

C. albicans has developed a successful relationship with man to the point where it is now inhabiting about 60 percent of humanity. Most human-to-human transfer occurs at the time of birth, but some is picked up later from close intimate contact. Oral sex, especially cunnilingus, is a factor in the spread of C. albicans.

Why Do Allergies Occur?

To try to find an analogy to explain how this pathogen interaction leads to immune dysfunction in the host animal is not an easy process. The fact that IgE is involved in both pathogen infections and allergies gives us a clue, but it does not explain causality. What we need is to be able to show how pathogens cause allergy. The best way to present an explanation is to look at C. albicans as penetrating barriers. The penetration of barriers was referred earlier in the abnormal bowel cascade. Invasive pathogens, such as worms and C. albicans, can invade the wall of the gut. This invasion stimulates a defensive response, and when all the prior defensive layers have been overcome, we recruit our final immune barrier of IgE. In this model, IgE is the final defense, and it acts as an alarm bell. When IgE is stimulated and causes mast cells to degranulate, other defense cells are directed to come to the site of degranulation. The mast cell degranulation is only useless as a defense when it is directed against something that is not dangerous such as grass, foods, or house dust mite (these we call allergens). If the reaction is directed against a pathogen, it is beneficial indeed, as it helps us defend the body. To understand why we sometimes develop a response to something that is not dangerous—an allergen—we must understand the big picture.

The Big Picture

Pathogens such as C. albicans are trying to invade the body, while our immune system is trying to prevent it entering. The body can detect pathogens, including C. albicans, with the help of phagocytes and other immune factors. If pathogen invasion persists, the body's defenses will increase. Ultimately the body may recruit antibodies in the IgE group, which is best to think of as the final immune defense.

Our phagocytic cells are always sampling the surroundings. They are scouring the environment looking for clues that might be of benefit in defending the body against infection. The phagocytes sample proteins and other molecules in the subcutaneous tissues of the mucous membranes. Phagocytes are sampling the environment and looking to find compounds against which the body can mount an immune response. The phagocytes then take these proteins to the lymphocytes, which may be stationed in our adenoids, tonsils, Peyer's patches, or other lymphatic glands. The lymphocytes then process these proteins, be they related to grass, house dust mite, food, or indeed C. albicans remnants. The lymphocyte may identify the protein as not being worthy of forming an immune response against or as foreign and worthy of an immune response (it is then called an antigen). If we react against the antigen with an immune response, be it T cell and delayed or immediate with antibodies, including IgE, we are mounting a defense. If the defense has been mounted against an antigen that is not dangerous, such as grass, house dust mite, or food we have developed an allergy, and the antigen is referred to as an allergen. The real intent was to create antibodies or an immune response against an antigen that would help us defend ourselves against C. albicans or other pathogens. In the case of C. albicans, this stealthy organism is difficult to defend against. The immune system can, in the process of defense, be tricked and identify the wrong culprit. When the immune response is directed against a non-dangerous allergen, such as cat hair or tomato, we have an allergy.

People are not all the same. People vary in race, sex, strength, etc. We also vary in how our immune systems are put together. Our immune system structure is dependent on our DNA. Our DNA gives us a programmed way in which we will tend to react to foreign invasion. Different individuals will react quite differently to the same infection. You have probably noticed that even within a family some individuals will tend to get similar symptoms to the same infection while others will not. What we have, therefore, is different immune responses depending on our genetic makeup. The immune system, which is genetically coded, directs how the body is to fight a given infection. With the same infection some people will get achiness, a runny nose, and sinus issues, while others might get a sore throat or no symptoms at all. Most people will accept that this is true from personal experience. As a survival strategy for humans, genetic variation is very important. If everyone reacted the same way, then a pandemic could wipe out everyone rather than just one third of the population as was the case with the black plague.

What is more controversial is that people with identical genetics (identical twins) may have different immune outcomes from an identical infection. How is this possible? It is possible because they do not have identical gut floras. The gut flora difference will influence the degree of invasion of the pathogen into our bodies. Those with an intact, well-working gut flora are less likely to develop allergies when exposed to C. albicans, while those with a damaged gut flora are more likely to develop allergies. This observation has always been troubling for medicine, which tends to think along the lines of one infective organism equals one disease outcome. What is apparent is that one infection, especially Candida albicans, can cause different diseases even in genetically identical individuals. The immune stimulation that results from the human/pathogen reaction causes the outcome. If the disease organism is not controlled, it will continue to stimulate the body's immune system, and further different disease presentations will occur over time. "Many children who outgrow their

AD (atopic dermatitis) develop other allergic diseases, such as rhinitis or asthma. The simultaneous development of allergic tolerance in one organ and the intolerance or atopic disease in another organ suggest that genetic, immunologic and environmental factors play a complex role in the natural history of AD and other atopic diseases" (Novembre, E., and A. Vierucci 2001, 105–108).

One organism can therefore result in many different immune-based diseases, which include not only allergies but also autoimmune diseases.

The huge array of symptoms that can be affected by allergy is staggering. Allergy can affect most organ systems. As medicine is divided into branches that are largely based on organ systems, it is not surprising to find that most branches of medicine are affected by one or more allergic diseases. The classic atopic allergic disorders are asthma, eczema (atopic dermatitis), and allergic rhinitis (nasal congestion). These three disorders are covered by three different branches of medicine: namely respiratory medicine, dermatology, and otolaryngology (ears, nose, and throat medicine). Allergies can cause abdominal pain or indigestion/reflux and affected patients may end up with surgeons or gastroenterologists. Allergic problems with children are seen by pediatricians, while psychiatric difficulties may see counsellors or psychiatrists. I cannot think of an organ system that cannot be affected by allergy or, for that matter, a branch of medicine that does not see allergies.

Allergy can change through time. A colicky baby may stop colic only to go on and develop eczema. With time the eczema too, may be lost, only for the child to develop asthma. Does this indicate that the dermatologist successfully treated the eczema and that the patient was cured? Not necessarily; it is likely that the same underlying disease process is still at work but that the organ affected has changed. Initially it was the gut that was affected and then the skin and now the lungs. This ability to change antibodies was discussed in the *Pediatric Allergy and Immunology* journal (Orivuori et al 2014, 329-337):

Increased IgA or IgG against BLG (Beta-lactoglobulin in milk) or gliadin (a protein in gluten) at age 1 was associated with IgE sensitization at age 6. We suggest that an enhanced antibody response to food antigens reflects mucosal tolerance aberrancies, e.g., altered microbiota and /or increased gut permeability, which is later seen as sensitization to allergens.

The fact that allergies can change from one antibody type to another and from one organ system to another organ partially explains why the area of allergy has been so difficult to study.

Generally, people do not become allergic to things they have not been exposed to. With peanut and bee sting allergy, the first exposure does nothing, but the second can be lethal. This area of sensitization has been studied extensively, and while some cross-reactivity of allergens can occur, it is not common.

In summary: When the gut flora is damaged, pathogen establishment and penetration is made easier. Pathogen attack stimulates our bodies' defenses. Initially, nonspecific cellular defenses of phagocytes are sufficient. If phagocytic defenses are insufficient to remove the pathogen, then immune stimulation will result. With prolonged pathogen infection, further immune defenses are stimulated. This may recruit immune mechanisms that are not advantageous, and this may result in what we know as allergy. Allergy is when a non-dangerous antigen, such as a tomato, grass, or house dust mite, is accidentally targeted by our immune system. We now refer to the antigen as being an allergen.

Nutrient Deficiencies

Some nutrient deficiencies, allergies, and most autoimmune diseases can be thought of as being "innocent bystander" diseases. The real battle is between the human body's immune system and the pathogen. This battle can spill over and damage the body. Nutrient deficiencies can be the result of damage to the gut lining by inflammation. Inflammation can result from allergies or the effects of the body trying to rid itself of pathogens. The gut doesn't care what causes the damage, as the result is the same. It is the size of the surface area of the gut that enables the body to absorb nutrition. Reducing the surface area decreases nutrient uptake, and malabsorption may result. The surface area can also be physically reduced by the direct effect of some pathogens. C. albicans produces adherent mycelial plaques on the gut lining that obstruct nutrient absorption.

The "leaky gut" may result from the absence of beneficial bacteria coating the gut lining, C. albicans, or other pathogens growing on the gut lining, or by damage to the gut wall due to inflammation. Food allergies directly damage the gut lining making it leaky when the offending food continues to be eaten.

Coeliac disease is one allergic disorder caused by gluten that is recognized by the medical profession as causing nutrient deficiency. When overgrowth or inflammation takes place in different areas of the gut, we can expect different nutritional deficiencies to result. Nutrients are absorbed in different areas of the gut but usually the small bowel. Fat soluble vitamins (vitamins A, D, E) are absorbed in the small bowel mostly in an area called the terminal ileum. Vitamin B12 is also absorbed here. If inflammation occurs here, *Keratosis pilaris* (bumpy skin on the upper arms and legs) may result from a vitamin A deficiency. B12 and vitamin D deficiencies can also occur when inflammation occurs in the terminal ileum. If inflammation occurs earlier in the ileum, an iron deficiency might result. A huge number of deficiencies can occur depending on where the inflammation and thus damage to the gut lining has occurred.

If the small bowel is "overgrown" by organisms, nutrient deficiencies are also expected. In some people high cholesterol levels are the result of small bowel overgrowth. Bile acid/cholesterol uncoupling can occur because of the organisms in the small bowel. The net effect is that the cholesterol is reabsorbed rather than excreted to the large bowel. A higher cholesterol level in the blood is the result.

Autoimmune Diseases and Causation

Autoimmune disease is acquired immunity to self that damages the body's own cells or cell functions. This can be either by direct effect of immune damage on cells (liver), or the result of an innocent bystander reaction (as in amyloid where organ structure is destroyed or when immune complexes are filtered by the kidney resulting in kidney damage) or by molecular mimicry (as in type 1 diabetes or thyroid disease).

Autoimmunity, like allergies, is the result of immune stimulation gone wrong. Except in cases of intracellular pathogens being removed by our immune system and thus killing tissue cells, autoimmunity is an immune reaction that is accidentally directed against cells of our own tissues. An innocent bystander reaction occurs when something is hurt by proximity to a battle that they were unlucky enough to be next to. By analogy it is when a peaceful wedding party in a restaurant has a party member hurt by an unruly patron who is involved in a brawl with another unruly patron. The innocent parties are inadvertently brought into the drama and can be hurt by an altercation they did not create and of which they did not want to be a part. A bottle might get thrown by an unruly patron, and this ends up missing the intended target and instead hits someone in the wedding party. That person is now the innocent bystander. By analogy our body organs can become innocent bystanders. A body organ is a victim of a battle going on elsewhere in the body. Most innocent bystander reactions result from disruption of the architecture of an organ. Amyloid disease lays down proteins in a tissue, disrupting its shape and making it no longer function. Immune complexes (a combination of an antibody and antigen) filtered by the kidney from the blood can lead to inflammation and damage to the kidney. This too is an unintended result and is described as an innocent bystander reaction.

Molecular mimicry is another type of autoimmune reaction resulting from immune dysfunction. Molecular mimicry occurs when a body protein is immunologically damaged or made to malfunction because it looks like a pathogen or presumed pathogen. In our wedding party/patron brawl analogy, molecular mimicry would occur when the police arrive and arrest all people with blue jackets, as the report had been made that the troublemakers wore blue jackets. Unfortunately for a person in the wedding party who had a blue jacket, it means that that person too is arrested by the police. Their arrest was unjust, but it occurred because the individual looked like the people who were involved in the brawl. In this situation a resemblance equals the crime. This is molecular mimicry.

Molecular Mimicry of Autoimmune

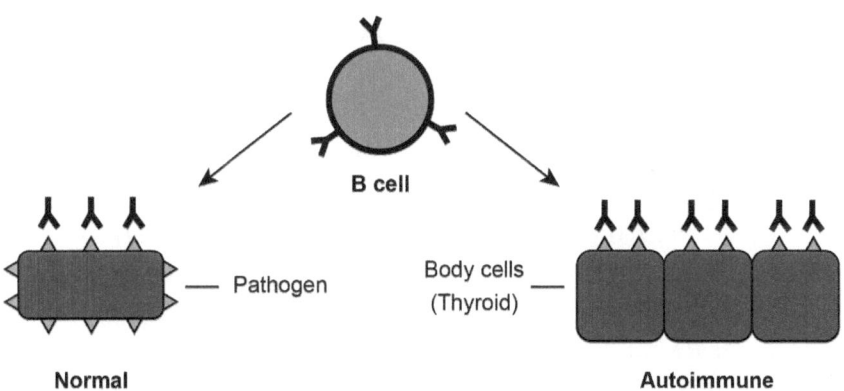

Thyroid disease, such as Hashimoto's thyroiditis, is such an example of molecular mimicry. Here, the thyroid cells that produce thyroxine are accidentally systematically destroyed. We develop antibodies to an environmental factor (pathogen), and these antibodies unfortunately have receptor sites that cross-react with not only the pathogen but also with our own thyroid tissues. When the antibodies are unable to distinguish between the pathogen and the thyroid tissue, damage ensues. In some people, an antibody will be produced against the pathogen that will cross-react to the receptor on thyroid cells. In this case the thyroid can be switched on, as is the case in Grave's disease. This is an accidental response that comes about because the proteins of the pathogen and thyroid receptor look very similar. The antibody not only attacks the real target, the pathogen, but it accidently also targets the thyroid receptor. This is what we call molecular mimicry. Unfortunately, in these cases antibodies are not able to distinguish between the pathogen and the thyroid cell (in Hashimoto's thyroiditis) or the receptor (in the case of Grave's disease), with the result that the thyroid is damaged or malfunctions. With time and damage to the thyroid cells, the thyroid in Hashimoto's thyroiditis is damaged to the point that it can no longer produce sufficient thyroid hormone for the needs of our body. At that point thyroid deficiency results and thyroid replacement therapy will have to commence. Ultimately if the process continues, the thyroid cells are totally destroyed, and hypothyroidism becomes complete and permanent.

It is worth mentioning that if the process of thyroid damage is picked up early enough and stopped by removing the pathogen, then the thyroid can often recover. Recovery is due to the presence of stem cells. Stem cells are in our blood, and they have the capacity to develop into other mature cell types. Stem cells can change into most tissues cells and provide the potential of repairing damaged tissues. If the pathogen that initiated the immune system damage is eliminated, then the tissue damage will no longer continue. Organ repair by stem cells may eventually reconstitute the previously damaged tissue. If the pathogen is not eliminated and the rate of damage is faster than the rate of repair possible by the stem cells, then the organ will ultimately fail.

Molecular mimicry may also develop antibodies to proteins. The A1 and A2 milk protein controversy is a case in point. Milk has a casein protein that can be classified as A1 or A2. The A1 casein milk protein looks like a protein on Beta cells of the pancreas (that produce insulin) in some type 1 diabetics. A2 milk protein does not look like pancreatic Beta cell protein. If the body has been stimulated to form antibodies to A1 casein, then these antibodies cross react with the Beta cells of the pancreas. The accidental destruction of the Beta cells results in the pancreas losing the ability to produce insulin. With no insulin production, type 1 diabetes ensues.

With the removal of A1 casein from milk it is hoped that type 1 diabetes will be avoided in susceptible individuals. Interestingly, NZ farmers are breeding out the A1 casein in favor of A2 casein. I commend this proactive behavior on the part of the NZ dairy industry.

The other way to prevent the development of diabetes is to remove the pathogen. The pathogen causes the body to create antibodies to the A1 milk protein. With the absence of the pathogen, no antibodies to the A1 protein are created, and no diabetes results.

Hypothetical Model for Causation of Type 1 Diabetes

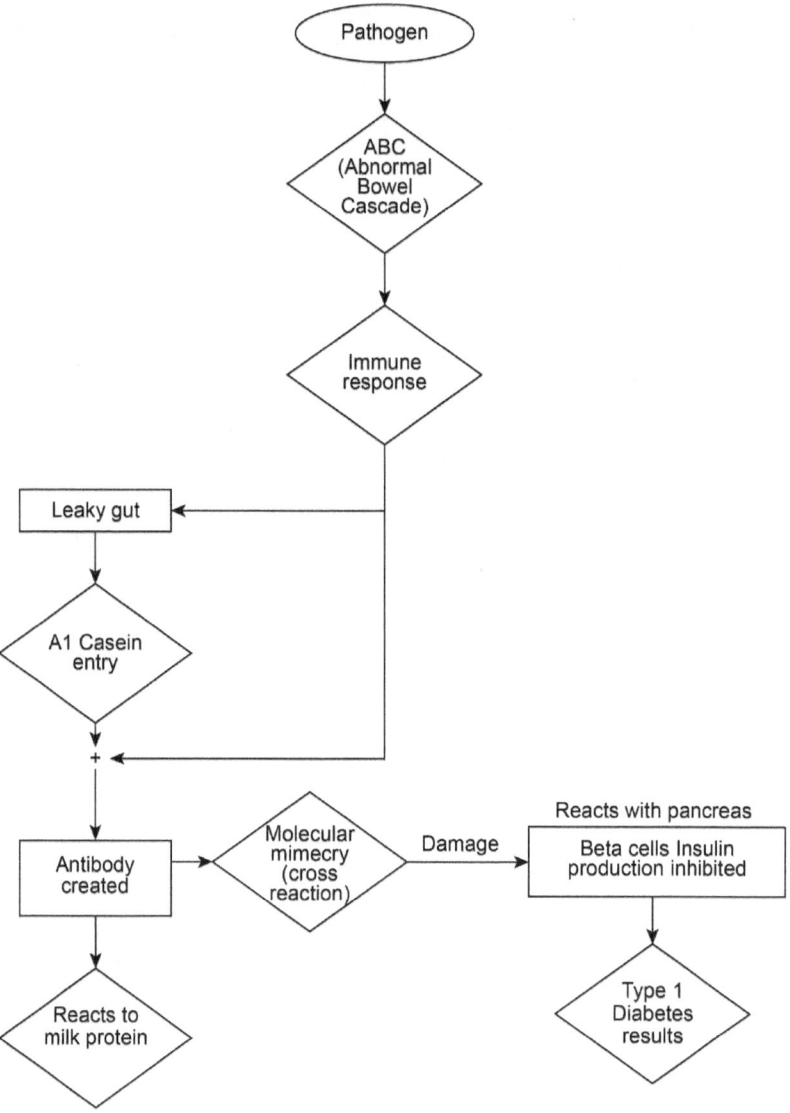

Dr. Neil B. Benson, MD

Treatment of Candida Albicans

If a healthy gut flora has not been established at birth, then problems commonly ensue. Constipation, diarrhea, bloating, and failure to thrive may result. A poor gut flora makes it easy for small amounts of candida to get established. Initially C. albicans will present as nappy/diaper rash, or it can be seen in the mouth or on the tongue. C. albicans may also be seen on the penis, labia, or skin folds. We might also see signs of impaired fungal immunity. While C. albicans does not directly cause cradle cap (seborrheic dermatitis and dandruff in adults), ringworm, tinea versicolor, or some types of athlete's foot, C. albicans is usually lurking in the background impairing fungal immunity. Removing the C. albicans generally cures the other fungal issues.

Thrush Direct Signs and Symptoms

Signs (what you see): diaper (nappy) rash, thrush in the mouth, anus, vagina, foreskin, groin (sweat rash), cheilitis (cracks at the corner of the mouth), premenstrual vaginal discharge.

Symptoms (what we complain of): Anal and vaginal itch, sweet tooth, nausea, or lack of hunger until late morning, premenstrual changes including depression, irritability, anger and anxiety, endometriosis, and others.

Associated signs and symptoms: Fungal impairment signs include cradle cap, dandruff, and other forms of seborrheic dermatitis, tinea versicolor, ringworm, athlete's foot, cracked heels, and nail infections.

If the C. albicans mycelium invades the gut lining, the affected child often develops allergies. Allergies can cause diarrhea, constipation, bloating, reflux, failure to thrive, eczema and many other symptoms.

If identified early on, C. albicans is often treatable by diet and by a single prescription medication called nystatin. What is nystatin, you say? In my opinion it is the second most useful drug of the twentieth century (second to penicillin).

Nystatin

(NY=New York, Stat=State, with the "e" dropped
and the "in" tacked on as it sounds nice).

Nystatin was discovered in a soil sample in a Virginia farm by two dedicated American scientists: microbiologist Elizabeth Hazen and chemist Rachel Brown (Baldwin 1981). Their discovery of nystatin in 1950 was to be the first antifungal antibiotic that proved effective and safe in the treatment of human disease. Both scientists worked at the Albany laboratory of the New York State Department of Health (now you know where the name came from). To answer one more common question, no, nystatin is not a "statin" and it has nothing to do with the group of medications used to treat cholesterol.

Nystatin

Chemical Structure

Nystatin A

Physical shape of Nystatin is circular.
* Diagram is adapted from Baldwin, Richard S., The Fungus Fighters Two Women Scientists and Their Discovery 1981 Cornell University Press Ltd, London, United Kingdom

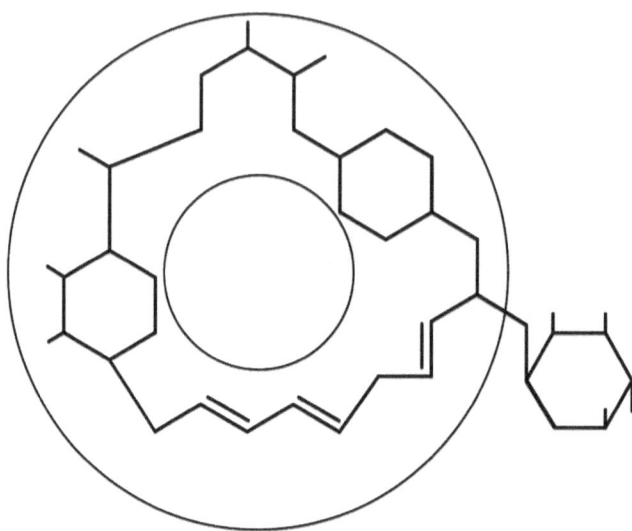

Nystatin fits into the wall of yeast causing a 'donut-like' hole to be created. The yeast cell contents are then leaked out of this hole.

What makes nystatin so useful is that the molecule is very large and essentially not absorbed across the gut lining. Most medications need to be absorbed into the body to work but not nystatin. Nystatin stays and works on the yeast form of C. albicans that is living on the lining of the gut. The medication does not interact with human cells, and it does not affect the bacteria in our gut flora. When we take nystatin orally it passes down the digestive tract and passes out in our feces. Passage will only be blocked if nystatin contacts C. albicans or food yeast. The nystatin will now penetrate the C. albicans or food yeast. For this reason, we should avoid eating food yeast at the same time as taking nystatin. If you continue to eat yeast, the medication is being wasted and is not available to kill C. albicans. Nystatin is probably the safest medication I know of, and it does not interact with any other medications of which I am aware.

Nystatin is a large doughnut-shaped molecule that fits into the wall of yeasts and creates a porthole effect that causes the yeast to leak its cell contents and then die. Yeast is the name given to a growth form of fungus.

This includes bread yeast or Saccharomyces cerevisiae, C. albicans, and many other fungi. When treating C. albicans, some people react to the leaked C. albicans yeast cell contents. This negative side effect is known as the Jarisch-Herxheimer reaction or the "die-off" reaction. This die-off reaction was originally described with penicillin. Penicillin weakens the wall of susceptible bacteria leading to the cells popping and releasing their cell contents. The popped bacteria have proteins and other cell compounds that when released can have a profound effect on the host patient. These compounds overwhelm the body in toxic compounds that make the patient initially look much worse.

The die-off reaction often occurs with C. albicans and nystatin. Die-off is more common in people with significant long-standing C. albicans colonies. Common symptoms are flu-like, with achiness, tiredness, and sometimes gut cramping. Other die-off symptoms can also occur, and these include skin rashes, worsening of eczema, arthritis flares, and many other symptoms.

Die Off Reaction

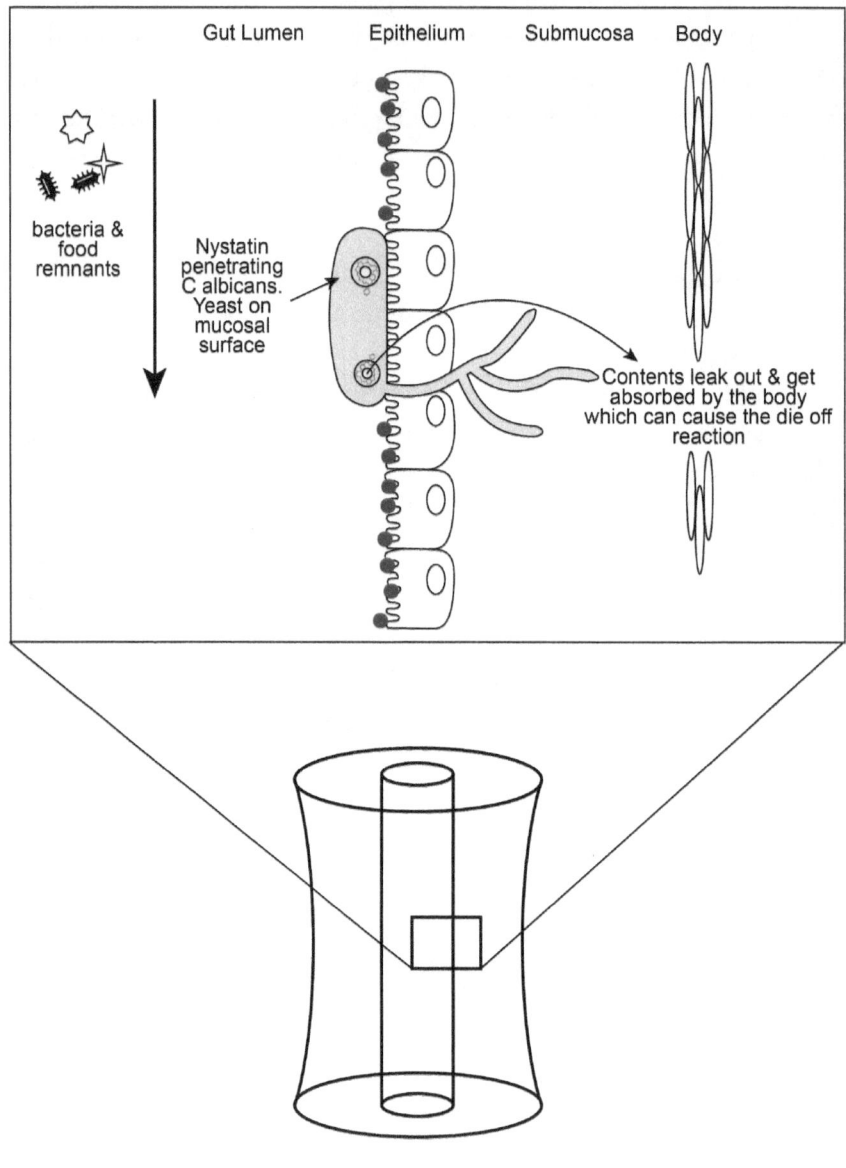

Dr. Neil B. Benson, MD

Unfortunately, as part of the process of improving health, the die-off process must be gone through. The effects can be minimized if the process is controlled. Usually this means starting at a low dose and increasing the dosage as symptoms allow. Not everyone gets the die-off reaction, but it can be very problematic for others. The important point to note is that the reaction is not directly related to the medication (and therefore not an allergic reaction) rather it is caused by the killing of the C. albicans. When the C. albicans is gone, the medication has no effect on the host. Some reports of an allergy to nystatin have been reported in the scientific literature. When studied more thoroughly, I noted that most of these "allergic" reactions relate to the die-off reaction. I am not saying that an allergy to nystatin cannot occur (more likely with topical sensitization), but I have not seen it occur in my career when nystatin is taken orally. An allergy to nystatin taken orally is very rare indeed. When I spoke to Dr. William G. Crook in 1995, he too had not seen an allergic reaction to oral nystatin.

C. albicans can live anywhere in the digestive tract. To ensure C. albicans is treated from the tip of the tongue and down throughout the gut, one capsule/tablet of nystatin should be opened onto the tongue or chewed in the mouth prior to swallowing. If you swallow capsules/tablets of nystatin, the medication will only take effect when the capsule or tablet dissolves in the stomach, and the first part of the gut will be missed. Let the medication stay in the mouth and disappear with your saliva. This will allow the nystatin to have time to work in the mouth. The bitter taste is hard for some people, but it will not hurt you, and I suggest you push through it, as the benefits outweigh the mild bad taste experienced.

Nystatin is effective at killing only the yeast phase on the surface of the gut mucosa. It cannot penetrate the wall of the gut and is not effective on the invasive mycelial stage of C. albicans. To treat the invasive phase a different medication must be used. I tend to use fluconazole (Diflucan is one well recognized brand name), which due to the long half-life in the body and depending on the dosage, means the medication can often be taken only every few days.

Fluconazole is systemically absorbed (unlike nystatin) and has interactions with some medications. It can have an unusual interaction with antihistamines, resulting in a rare heart arrhythmia called Torsade de pointes. This reaction is reported as being rare, and I personally have not seen it. Despite the rare nature of this adverse interaction, I believe it is good practice to avoid antihistamines if taking this medication. Fluconazole does not usually produce a die-off reaction but does have other side effects. It should be avoided in pregnancy or if severe liver disease is present. See your doctor prior to using either of these medications, but special care should be taken with fluconazole, which can interact with other medications.

The combination of nystatin killing the mucosal yeast form of candida and the fluconazole killing the invasive mycelial form has proven to be very effective if taken together. Start with nystatin alone to ensure the patient does not have a significant die-off reaction. Then sequentially increase the nystatin as tolerated by the patient. For adults that is until they are taking two capsules/tablets of nystatin three times a day (for a total of six per day, three opened in the mouth and three swallowed). It is only then that the fluconazole should be introduced.

Fluconazole can be given every third day due to the medication having a long half-life (the time for half the medication to be inactivated or excreted), for adults use 200 mg every third day. This is not a high dosage, but it is effective.

A usual starting course for these medications is fifty capsules of nystatin on its own, and depending on how the die-off reaction has gone, the nystatin will be continued at six per day accompanied by one 200 mg fluconazole every third day in combination thereafter. A usual course of fluconazole is only three capsules, one every third day.

In mild cases of colonization, C. albicans can sometimes be eliminated with one course of antifungal antibiotics. For tougher cases more medication courses will be required. Long courses of continuous antifungals are not generally recommended; pulsed courses followed by rest periods are probably ideal. C. albicans has resistant spore-like phases that can lock together in protective balls that make the resistant phase less sensitive to the medications. Episodic dosages are more likely to catch C. albicans in an active, medication-sensitive phase. Such an approach will waste less medication, reduce the risk of drug resistance, and increase the efficacy. Regular follow-up after a short course of treatment is recommended to check and make sure that the patient is eating a good diet and avoiding his or her known allergies. It is also a time to recheck on how the patient is doing and to consider other previously undiagnosed health issues or allergies as the reason treatment has not been successful. There will always be variation in treatments depending on the patient.

Some women who have not had thrush for many years or even decades will find that after their initial course or courses of antifungals they develop thrush. This is common in women who are overwhelmed by C. albicans. While it is counterintuitive that thrush should occur after treatment when it was not apparent before treatment, it can be explained on the grounds of recognizing that when the body is overwhelmed it is not able to mount a sufficient defense. On knocking back the C. albicans with the antifungals the body's immunity improves, and the woman shows symptoms of vaginal thrush with discharge and itch. The antifungals did not cause the thrush; rather, the body is now working better and is able to put up a defense and thus the symptoms arise. Continuing the antifungals will eventually get rid of the thrush, and then the symptoms abate.

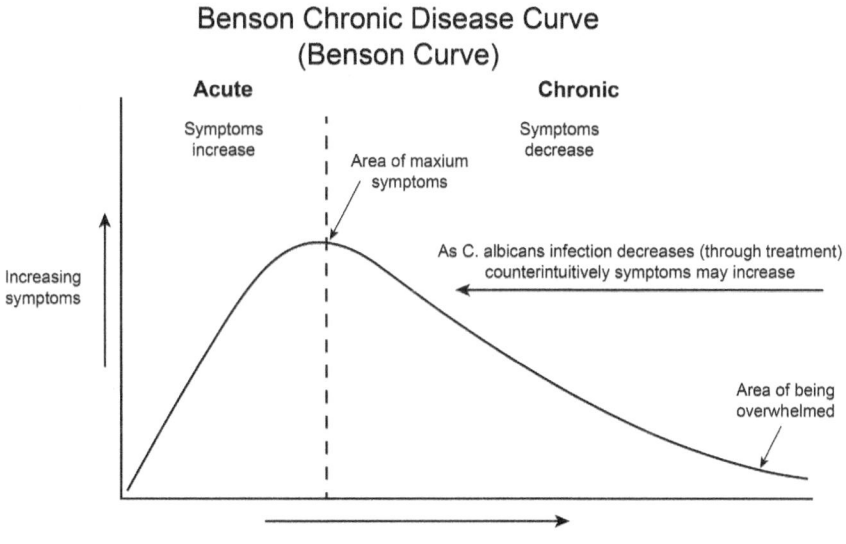

Benson Chronic Disease Curve
(Benson Curve)

Acute | Chronic

Symptoms increase

Area of maxium symptoms

Symptoms decrease

As C. albicans infection decreases (through treatment) counterintuitively symptoms may increase

Increasing symptoms

Area of being overwhelmed

Increasing Infection with C. albicans

The Benson curve is useful in this discussion of chronic infection with C. albicans and in chronic infections generally. When our body is in the chronic phase and the immune response has been maximized, we no longer have the immune reserves to fight off new invaders and are vulnerable to getting other infections. This is the case clinically with children who have a C. albicans infection. These children are much more likely to have additional infections, such as common warts, molluscum contagiosum, ringworm, athlete's foot, seborrheic dermatitis, cold sores, and many other infections. If they get chicken pox or shingles, their outbreaks tend to be worse than in children not affected by C. albicans. If additional infections are encountered, they will last longer, and the child is likely to be known as a child who "catches everything that is going around" or is "sick all the time." If treated for the C. albicans the child becomes more resilient—able to overcome infections more easily and with fewer symptoms that are less severe and have shorter sickness periods.

When other conditions are present in addition to C. albicans, and C. albicans is treated, I have commonly observed that previously difficult-to-control disorders become controllable. Gout and diabetes are two disorders that can greatly benefit from C. albicans treatment. Following treatment for C. albicans, I have also noticed that sun damaged skin, liver spots, and warts will often disappear on their own, without any treatment being addressed specifically to the skin disorder. The improved immunity has now been able to address health issues that it was not capable of combating when the immunity was overwhelmed. When we are on the chronic side of the graph and have an overwhelmed immune system, we are much more likely to suffer from additional infections and disorders. Treating the C. albicans will have the benefit of improving our health and immunity generally.

I am sure that other chronic infections also adhere to this observation model of immune compromise. By clearing the one infection, the second problem becomes easier to manage. When encountering a difficult clinical situation, it is always wise to consider additional chronic infections, as they may have a role to play. This area needs further study.

Natural Antifungals and a Warning

Nature produces a variety of natural antifungals. The antifungals are usually found in plants. Many herbs, and especially oregano, are potent antifungals. Caprylic acid from coconut, olive oil and olive leaf extract, garlic, pau d'arco, licorice root, and many other products are also antifungals. In my experience nothing works as well as nystatin and fluconazole used together. Adding some of these natural antifungals, especially garlic and herbs to a good diet, will provide additional antifungal benefits. Studies need to be done on natural antifungals to ensure compatibility with traditional medications and to ensure they do not damage the gut flora.

Many people are trying to improve their health by changing their diet. While this is commendable, there are some pitfalls that should be avoided. Changing to a simple diet, whether that be a carnivore (meat only) diet or some other diet, will often make people feel better in the short term. Usually this dietary change removes what they have been allergic to with the result that they temporarily feel better. With time if the underlying causative pathogen has not been dealt with, the person will ultimately find himself or herself in a worse situation. I made this mistake and became allergic to essentially all foods. I had not removed the C. albicans sufficiently and had not identified a significant beef allergy, and through time I noticed my nonallergic food range became smaller and smaller. I recommend you treat thrush thoroughly at the start, in addition to allergy avoidance, and this should help prevent the patient becoming "allergic to everything."

All about Allergies

Allergies can be investigated by skin-prick testing and blood tests known as Radioallergosorbent testing (RAST test). These tests are typically used for IgE antibodies. Both tests are equally reliable, and both require some interpretation. If a test is positive, it says there is a problem. If it is negative, it does not infer that the patient does not have a problem with that allergen, be it a food or environmental allergen. The confusion here is because we are only testing IgE. Other antibodies can also be involved, and these are often not tested for. IgA and IgG or even cell-mediated reactions might be involved, and sadly we are usually unable to check for these specific antibodies and reactions. Hopefully labs will develop a more complete range of tests in the future.

The incomplete nature of testing was discussed in the *European Journal of Clinical Nutrition* when it discussed cow's milk protein allergy (CMPA) and cow's milk protein intolerance (CMPI). The article "The natural history of cow's milk protein allergy/intolerance" (Høst et al 1995, 13-18) pointed out that "adverse reactions to cow's milk protein(s) may be due to the interaction between one or more milk proteins and one or more immune mechanisms, possibly any of the four basic types of hypersensitivity reactions.

"Thus, the classification of adverse reactions to cow's milk proteins depends on the extent and quality of performed diagnostic test for immune mediated reactions" (Høst et al 1995, 13-18). They go on to say that "at present, no single laboratory test is diagnostic for CMPA/CMPI, and differentiation between CMPA and CMPI cannot be based solely on clinical symptoms." This situation today is still the same as in 1995 and "the diagnosis has to be based on strict well-defined elimination and milk challenge procedure" (Høst et al 1995, 13-18). The gold standard will always be avoidance and rechallenge, as we cannot rely on negative laboratory tests predicting a negative immune reaction because of the incomplete nature of the laboratory testing procedure.

I noticed that with skin-prick testing you can see additional immune-based reactions if the skin-prick sites are watched over twenty-four hours. Skin reactions will not only come up within twenty minutes; they will come and go over time, and some will even be detectable days later. We normally only watch the sites for half an hour, which is ample for IgE-mediated reactions, but we are in fact missing many other immune-based reactions that occur later. If you are keen and are skin tested, see if you can notice other reactions that come and go after the initial twenty minutes.

Food and environmental allergies develop from exposure. It is unlikely you will develop an allergy without previous exposure. Some rare in utero sensitization can occur and some cross-sensitivities have been reported. Environmental allergies include house dust mites, grass, tree and plant pollen, and pets. Food allergies are also related to exposure. A corn allergy is much more common in North America. Corn and the derivative sweetener corn syrup are commonly used in foods in North America. In Australia corn allergies are less common. Sugar cane is the main sugar used in Australia and corn is less commonly encountered in the diet. Corn allergies do occur in Australia, but they are less common. Sugar cane allergy occurs occasionally in Australia, but it rarely occurs in North America. Food allergies can develop to any food, and all too often it is the person's favorite food, which is usually the result of frequent eating of that food.

Common symptoms of allergy can be related to the gut. When the bowel is confronted by something offensive, the bowel will move the offensive thing out of the gut as quickly as possible. This may be by causing vomiting or diarrhea or may present in milder forms as heartburn/reflux or loose, splattered bowel motions. This symptom of food allergy has become very common today with people describing this observation to coffee as being "coffee makes me poop" or to alcohol as being "booze poops."

The increase in allergy was discussed in *Immunology Letters* (Niederberger 2009, 131–133) when he pointed out that "in 1906 less than 1% of the population were affected" while "today, more than 25% of the inhabitants of industrialized countries suffer from allergic rhinitis, conjunctivitis, asthma, dermatitis or food allergies." While this increase is massive, I believe it is more like 50 percent of the population that suffers.

Mold allergies are an unusual group. Many mold allergies are the result of environmental exposure to either a moldy environment or moldy foods. The presence of an antibody may also indicate the mold is living in the body. A person can have IgE antibodies to a range of our environmental molds, including some pathogens, such as Aspergillus in lungs and C. albicans in the gut. Antibodies to C. albicans are the result of being colonized by C. albicans. If antibodies to C. albicans are present, then care must be taken when treating with nystatin, as a die-off reaction commonly results. We probably should not look at IgE antibodies to C. albicans as being an allergy. In this case antibodies to C. albicans are not an inappropriate allergic response but a normal immune defense to C. albicans.

A food yeast allergy is frequent and is typically related to a C. albicans gut infection. The yeast form is common to both organisms. Cross-reacting to the surface of the candida and Saccharomyces cerevisiae (bread and brewer's yeast) fools our immune system into reacting to both. I do not recommend taking a probiotic with food yeast in it, as food yeast does not normally live in the human gut. Further, eating food yeast can waste our immune reserves. Cross-reactivity of yeast and C. albicans is common. If cross-reactivity occurs, we waste the immune response fighting the food yeast that should be available for fighting C. albicans.

Dr. Neil B. Benson, MD

Avoidance and Rechallenge

When a food allergy is suspected or if it has come up positive on skin or blood test, there is an approach that can be used to see how it affects you. We call this process avoidance and rechallenge. The process involves avoiding a suspected food for five days and then on day six having it again. With this approach we expect that the symptoms should improve when off the food and for the symptoms to return in an exaggerated form on reintroduction. The increased symptoms on rechallenge are the result of the mast cells releasing more allergic mediators. With avoidance the mediators were not used and were stored up. On rechallenge with the allergen, the mast cell now releases the larger store of mediators, giving an amplified response.

When people have deliberately given up a food that is causing them symptoms, it is common that on accidental reintroduction they will have exaggerated symptoms. Fortunately, with time these symptoms decline. When no longer challenging the body with the allergen, the stimulation of lymphocytes stops, and the production of antibodies declines. The net result of avoidance is a fall in available antibodies over time. If left alone for long enough, the immune system will eventually stop producing these antibodies and forget the allergy. You have then lost the allergy, and you are cured. This can take years to occur and longer if avoidance has not been complete, and stimulation has persisted. In all cases the pathogen must also be removed for cure to ultimately occur.

A term tolerance is found in the literature, and this describes how some people lose their allergic reaction over time while continuing to have exposure to the allergen. It is suggested that the body turns off the allergic response, and while this might be the case, I am not aware of how exactly it works. It is this created tolerance that is also sought in immunotherapy (allergy shots).

Allergy Shots

Most traditional allergists today take the approach of desensitizing with injections. While desensitization can be quite successful, it does have a downside. Anaphylactic reactions and even death sometimes occur. The anaphylactic reaction can be immediate or delayed by hours. Patients must be made aware of this. The technique does not appear to work for food, and the effect with molds can be unpredictable. Desensitizing to C. albicans is probably not a good idea, as we are blocking a part of the body's natural defense against this organism.

Oral desensitization is now available for some environmental allergens. This approach is promising and may have less chance of inducing an anaphylactic reaction. The downside is that it is presently very expensive.

Other techniques to reduce allergy symptoms have been tried over the years, and they include enzyme potentiated desensitization. These nontraditional techniques appear not to work.

How to Test for Allergies

So, what things can be tested on skin and blood tests?

We test for what is commonly available through laboratories. This is an incomplete picture. IgE, which is what is being assessed, is not the only antibody or immune reaction that can be involved in allergic reactions. IgA and IgG antibodies are often involved, but they cannot be routinely tested for in most community labs.

The common environmental allergies include house dust mites, grass, trees, flowers, pets, and animals. Wasps and bees can also be tested for, but they should be done by blood testing. Skin-prick test should be avoided for wasps and bees, as it is potentially dangerous and can set off an anaphylactic reaction.

Mold patches can also be tested for common molds including alternaria, fusarium, and aspergillus. These environmental allergens are also commonly found in food spoilage. Aspergillus and other molds can also live in the lungs. There are also many other molds, but the most interesting is Candida albicans. If you have an allergy to C. albicans, you are presently or have been colonized by C. albicans. In this case the reaction is not a maladaptive, allergic reaction; rather, it is protective.

One thing to be aware of if you have a yeast or mold allergy is to avoid commercial fruit juices, commercial tomato products, and dried fruits including raisins and currants. The best fruit and tomatoes go to the market and the subprime fruit, some of which are moldy, are made into juices and sauces. A bunch of grapes often contains a few spoiled fruits.

When raisins are made, the moldy fruit are dried right along with the rest of the grapes, and this can make you sick.

Beans can be a problem when it comes to molds. Molds can colonize beans in the pod, and when they are dried, the mold remains. Sorting and removing the wrinkled beans, which signifies mold damage, used to be the job of the senior ladies in traditional societies. Today beans are rarely sorted with the result that mold-affected individuals may find that beans are a source of allergy for them.

Food

You can become allergic to any food, but the common things we eat are the things that should be tested. Wheat, milk, peanuts, corn, eggs (white and yolk are separate allergens), fish, shellfish (oysters, clams etc.), crustaceans (lobster, crab, shrimp, etc.), soy, and food yeast (Saccharomyces cerevisiae) are all common allergens.

The testing of wheat and yeast are more complete than most other allergens, as we have tests for the three main groups of antibodies. IgE can be tested for as well as IgA and IgG. Ask for the IgE antibodies for both wheat and yeast via RAST testing. For wheat also ask for coeliac antibodies known as TTG (Tissue transglutaminase antibodies) while for yeast we look at ASCA (anti-saccharomyces cerevisiae antibodies). ASCA is a test commonly used in the classification of inflammatory bowel disease (IBD), but it also tells us if antibodies to yeast are present. It is interesting that 66 percent of people with Crohn's disease have yeast antibodies (Campbell 2014, 75)! I have found that people with ASCA antibodies benefit from yeast avoidance.

Yeast allergy, S. cerevisiae, is one of the most overlooked allergies, and yeast avoidance is difficult. Yeast is often not labeled as an ingredient. Many things have yeast in them. Usually fermented means S. cerevisiae fermentation. There are some exceptions (sourdough, kimchi, and sauerkraut being chief among them). Fermented alcohols include beer, wine, port, sherry, and cider (Spirits which are distilled are yeast free). Apple cider vinegar and wine vinegar are fermented, but cheap white vinegar is generally okay, as it is chemically synthesized and not fermented. Yeast spreads include Vegemite, Marmite, and others, and these should be avoided. Bread has yeast (real sourdough does not have S. cerevisiae in it, but may have other yeasts) and fermented drinks such as kombucha have yeast and can include S. cerevisiae among the bacteria. Commercial dried fruits such as raisins and currants should be avoided, as should things that use yeast as a flavoring agent, and this includes commercial gravy mixes. Yeast may also be described as yeast extract or even vegetable protein. Be aware and read the packaging, as it will surprise you how commonly yeast is added as a flavoring to products. Who would have thought that salt and vinegar potato chips have yeast in them?

Yeast allergies are one of the harder allergies to avoid, but with perseverance it will be overcome. Thankfully the world is adapting to different diets; for example, some pizza restaurants now have sourdough bases as an option. If you look you shall find, and if you don't find, then ask. Successful businesses cater to their customers in need, and this brings about change.

Yeast Avoidance and Rechallenge

Yeast rechallenge once again involves avoidance for five days and on day six reintroduction. Here we use bread yeast S. cerevisiae (yes, from the bread-making aisle in the grocery store). Rechallenge initially involves having one teaspoon taken orally, followed by a glass of water. Then wait twenty-four hours to see the result. If *no* result is observed, then the next day do a second larger rechallenge. On the second rechallenge, a tablespoon of yeast should be taken. Once again wait and watch for the next twenty-four hours to see the effect.

Many patients will describe fascinating symptoms with food avoidance and rechallenges. Many symptoms are possible. The patient might find that with avoidance the nose clears and achy legs and shoulders disappear or that mental alertness and clarity improves. With reintroduction the previous symptoms recur, and the patient is made aware of how the food affects them. Sometimes the symptoms are subtle. This is the case when sleep is disturbed. The result may also be quite surprising and very apparent when a person gets agitated, anxious, aggressive or suffers from bladder irritation, bed-wetting, or acne. Whatever the response is, it will be repeatable, and I believe that all moderate or greater food allergies should be considered for a rechallenge. If anaphylaxis is expected, a rechallenge should not be done. Significant peanut allergies should never be rechallenged outside of a medical facility, nor should allergens that have given anaphylaxis previously. Care must also be taken with allergens that are suspected to cause asthma. It is advisable to consult with a medical professional before rechallenge.

If the blood/skin testing is positive, then it means you do have a problem with the allergen. If the avoidance and rechallenge is negative despite the positive blood/skin test, this may be because the result was incorrectly reported in the first place. In such a case, repeat the test. It could also be that the effect of the rechallenge was missed. It is not unusual for the rechallenge to affect areas we had not considered. It may affect a body organ you had not noticed such as the lungs causing asthma, or other organs that cannot easily be checked, such as the kidneys.

Mental health is also frequently affected. The emotions of irritability, anger, and anxiety are commonly affected. Euphoria is less common, and even sleep can be affected. A patient has even described how they have had auditory hallucinations (hearing voices) with rechallenge, while others describe being paranoid, stoned, and even hyperactive with racing thoughts, as if they were on amphetamines. Still others will suffer from brain fog. A patient who had been a psychiatric nurse once said to me, "Who would believe that eating fish would cause voices in my head? I wonder how many of my previous schizophrenic patients have food allergies?" If you are rechallenging yourself, make others aware of the test so that they too can comment on your behavior, as they may notice things you have missed.

The effect of allergies as the causal agent in some children's autism should be investigated. If allergy or C. albicans is the cause, how sad that we have missed this diagnosis, which could have so vastly affected change in the life of the individual and his or her family.

In food allergy testing, if these tests show you have antibodies to an antigen/allergen, then you have a problem with that antigen/allergen. If the test is negative, it unfortunately does not rule out that you can have a problem with the food allergen. When the allergy test is negative but you feel you have a problem with the allergen, then always consider doing avoidance and rechallenge, as this is the "gold standard" and most reliable test of all. The avoidance and rechallenge will pick up reactions that have been missed due to the incomplete nature of the skin testing and blood testing due to our inability to test all antibodies and immune responses.

In summary: Allergy testing can be very useful. There are subtleties with interpretation and limitations to allergy testing. Most allergy testing involves only IgE, and this gives an incomplete picture. Other antibodies and immune reactions can also be involved. Many if not all organ systems can be affected by allergy. The variety of presentations is truly staggering, and avoidance and rechallenge is the "gold standard" and ultimate arbiter of negative laboratory results. Positive results mean that the person is affected by the allergen, but it is not always apparent how the person is affected due to our inability to note reactions in some affected organs. A negative blood or skin reaction, which tests IgE only, does not mean the person is safe with the allergen. With negative results, avoidance and rechallenge should be considered, and if safe to do so, conducted.

To summarize many of the factors discussed in this book, I have included a flow chart that shows how pathogen persistence is central to allergic and autoimmune disease. Ultimately, I hope this book explains to you the origin of chronic inflammatory diseases.

For the scientists and clinicians, I recommend the following simple approach to achieve publishable results. Chronic diseases are multifactorial in causation, and this makes treatment difficult to study. For chronic diseases, the three variables to be studied are diet, allergies, and pathogens. Start with a comprehensive history sheet of signs and symptoms. This sign/symptom sheet should be completed at the start of each presentation. Presentations will include an initial consultation, which identifies the pretreatment status, including diet, and looks at factors requiring investigation. The first consult goes on to discuss the healthy plate diet and requests that patient-specific blood tests be obtained. A second consultation looks at the effect of adherence to the healthy plate diet and concludes with a discussion of the blood results and the patient's specific allergens. Avoidance and possible rechallenge strategies are then discussed. The third consult may extend to several attendances and assesses the effect of avoidance and rechallenge with the identified allergens and concludes with a medical treatment recommendation for the underlying pathogen or pathogens. The fourth and final consults are to assesses the effect of medical treatment. In clinical practice I use this four-part approach. It allows for each intervention to be separately quantified as a causal factor contributing to the overall health of the patient. Such an approach, properly applied, yields publishable results.

Pathogen persistance needed for causation of allergy and autoimmune disease

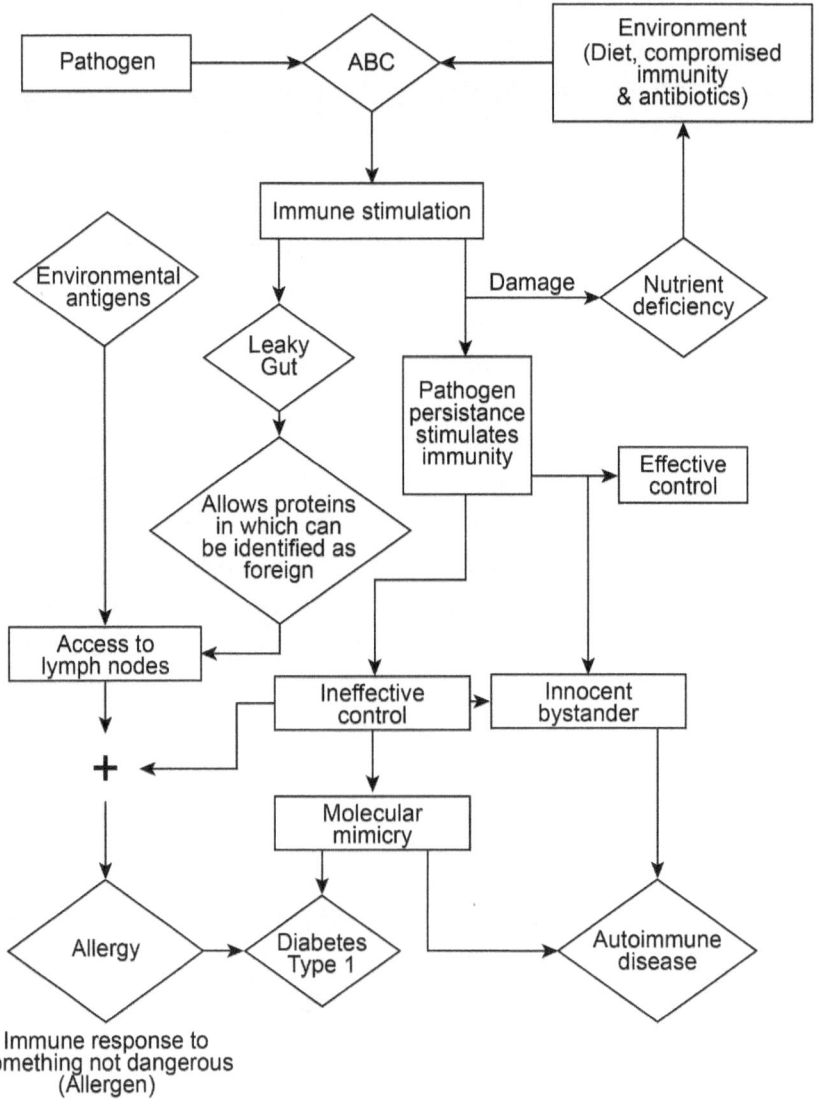

Dr. Neil B. Benson, MD

How Is My Health Today?

I started this book describing my poor health, and I am pleased to report that over time and with ups and downs, eventually my health dramatically improved. I do not take any regular medications and did not have the once recommended third nasal surgery. I can breathe through my nose except when I have a cold. With colds I also sometimes need to use an asthma inhaler. On increasingly rare occasions, I still take antifungals medications, but generally I am in good health, and I have thrush under control. I still largely avoid corn and yeast, but my other food allergies are all essentially gone. I have gone from being allergic to almost all foods to being almost free from allergies. I feel healthier than ever before. I encourage you to work hard on your allergies. Always try to eat a good diet high in green vegetables and encourage your friends and family to do this too. I wish you good health and hope you have found this information helpful. If you have benefitted from the book, please let other people know about it, so that together we can make a healthier and happier world.

Best wishes and kind regards,

Dr. Neil Benson MD, FRACGP, FRNZCGP

12 September 1994

Dr N Benson
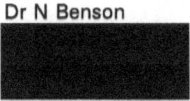

Dear Neil

I have now had the opportunity of discussing your proposal in depth with ███████ ███████ and ████████████, and need to report as follows.

Without being too cynical, I think you are aware that the pharmaceutical industry, and our two major client groups, i.e the conventional medical practitioner and the retail pharmacist, are basically not interested in cure of illness or disease. In a free enterprise situation all parties are basically interested in repeat business and return and protection of investment, whether it is to their own personnel interest or, in the case of the industry, to the shareholders.

In this regard, it is very difficult to step outside the conventions of supplying and developing treatments - particularly for chronic disease - which means on-going business; or invest in cure - which in theory is self-limiting. As we discussed, the helicobacter story is a good example of this.

What you are proposing to us is, to our mind, a system of therapy. A system that sounds exciting, and if followed should increase the "cure rate" for a large number of patients suffering from asthma and many diseases where there is an allergic component.

Our problem is how to package a system of treatment into a presentation that can be developed, registered, and marketed. Unfortunately we see major problems, unsurmountable to the state of knowledge that we hold, and the industry in which we participate.

Time has also proven that the regulatory requirements in product development will not be waived by regulatory bodies to enable a low cost cure to be developed. In fact, regulatory control and complexity, and cost, is becoming a dominating factor in any business development of this type.

In short, we do not think that this company can take your project any further. The only suggestion that we can make for you is that you should explore the opportunities to franchise the treatment system, or alternatively develop a clinic system under your own control where effectively you move to a chain of diagnostic and therapeutic outlets. This is probably contrary to the traditions of medicine, and you would anticipate considerable opposition - as happened when certain individuals tried to develop general practice into a chain-store type operation here some years ago.

However, we do not see a workable commercial alternative.

Neil, we very much enjoyed your visit and the fascinating discussion which eventuated, however are pessimistic that you will gain further support from the conventional pharmaceutical industry.

Yours sincerely

References

Baldwin, Richard S. 1981. *The Fungus Fighters Two Women Scientists and Their Discovery.* 1st ed. London, United Kingdom: Cornell University Press Ltd.

Benson, N. B. 1995. "M.E. and Allergy: The Abnormal Bowel Cascade" (Lecture). *Chronic Fatigue Symposium.* Recorded on 10-12 February 1995, Dunedin, Otago, New Zealand: The University of Otago. Video Format .mxf (Media Exchange Format). 30 minutes.

Campbell, K., ed. 2014. *Sonic Pathology Handbook: A Guide to Interpretation of Pathology Tests.* 1st ed. Australia: Sonic Healthcare Ltd, The Buckner Group.

Høst, A. Jacobsen, HP. Halken, S. and Holmenlund, D. 1995. "The natural history of cow's milk protein allergy/intolerance." *European Journal of Clinical Nutrition* 49, no. 1 (September): 13-18.

Huber, B. 2006. "100 years of allergy: Clemens von Pirquet – his idea of allergy and its immanent concept of disease (English Abstract)." *Wiener Klinische Wochenschrift* 118, no. 19-20 (October): 573-579.

Igea, J. M. 2013. "The history of the idea of allergy." *Allergy* 68, no. 8 (August): 966-973.

Kellock, B. 1985. *The Fibre Man: The life-story of Dr Denis Burkitt.* 1st ed. Tring, Herts, England: Lion Publishing plc.

Kuhn, Thomas S. 1962. "The structure of scientific revolutions." *Chicago and London.*

Marshall, Barry J. 2005. "Barry J. Marshall – Biographical." *NobelPrize. org*, 2005. Accessed: Mon. 13 Jan 2020. <https://www.nobelprize.org/prizes/medicine/2005/marshall/biographical/> "When the work was presented, my results were disputed and disbelieved, not on the basis of science but because they simply could not be true. It was often said that no one was able to replicate my results. This was untrue but became part of the folklore of the period. I was told that the bacteria were either contaminants or harmless commensals."

Mayr, E. 1982. *The Growth of Biological Thought; Diversity, Evolution, and Inheritance.* 1st ed. Cambridge, Massachusetts: The Belknap Press of Harvard University Press.

Niederberger, V. 2009. "Allergen-specific immunotherapy." *Immunology Letters* 122, no. 2 (February): 131-133.

Novembre, E., and A. Vierucci. 2001. "Milk allergy/intolerance and atopic dermatitis in infancy and childhood." *Allergy* 56, no. s67 (April): 105-108.

O'Hanlon, D.E., Come, R.A., and Moench, T.R. 2019. "Vaginal pH measured in vivo: lactobacilli determine pH and lactic acid concentration." *BMC Microbiology* 19, no. 13 (January): doi:10.1186/s12866-019-1388-8.

Orivuori, Laura, Kirsi Mustonen, Caroline Roduit, Charlotte Braun-Fahrländer, Jean-Charles Dalphin, Jon Genuneit, Roger Lauener et al. "Immunoglobulin A and immunoglobulin G antibodies against β-lactoglobulin and gliadin at age 1 associate with immunoglobulin E sensitization at age 6." *Pediatric Allergy and Immunology* 25, no. 4 (2014): 329-337.

Other Reading:

Crook, W. G. 1986. *The Yeast Connection A Medical Breakthrough.* 1st ed. New York, New York: Vintage Books (A division of Random House).

Crook, W. G. 1992. *Chronic Fatigue Syndrome and the Yeast Connection.* 1st ed. Jackson, Tennessee: Professional Books, Inc.

Crook, W. G. 1995. *The Yeast Connection and the Women.* 1st ed. Jackson, Tennessee: Professional Books, Inc.

About the Author

Dr. Benson has a strong Science background with a BSc. in Botany and a BSc. Agriculture. Graduating with a MD degree from the University of British Columbia in Canada in 1988, Dr Benson has gone to work as a GP in not only Canada but also New Zealand and finally Australia. He is the proud father of four grown children.